Aluminum Recycling

SECOND EDITION

Aluminum
Recycling
SECOND EDITION

Mark E. Schlesinger

CRC Press
Taylor & Francis Group
Boca Raton London New York

CRC Press is an imprint of the
Taylor & Francis Group, an **informa** business

CRC Press
Taylor & Francis Group
6000 Broken Sound Parkway NW, Suite 300
Boca Raton, FL 33487-2742

First issued in paperback 2017

Version Date: 20140407

ISBN 13: 978-1-4665-7024-5 (hbk)
ISBN 13: 978-1-138-07304-3 (pbk)

Library of Congress Cataloging-in-Publication Data

Schlesinger, Mark E.
    Aluminum recycling / author, Mark E. Schlesinger. -- Second edition.
        pages cm
    Includes bibliographical references and index.
    ISBN 978-1-4665-7024-5 (hardback)
    1. Aluminum--Recycling. 2. Scrap metals--Recycling. 3. Extraction (Chemistry) 4. Aluminum--Metallurgy. I. Title.

    TD812.5.A48S35 2014
    673'.7220286--dc23   2013045725

Visit the Taylor & Francis Web site at
http://www.taylorandfrancis.com

and the CRC Press Web site at
http://www.crcpress.com

Contents

Preface

When I wrote the first edition of *Aluminum Recycling,* I did so because I could find nothing like it. Seven years later, this has changed. Two excellent books, dozens of conferences and workshops, numerous reviews, and a Wikipedia entry have made it clear that this is a topic whose time has come!

However, much of what I have seen makes it clear that there is still a need for a clear perspective of the recycling industry and the place that recycling has in the production and use of aluminum by our society. Recycling is too often viewed in near-religious terms and not seen as a business that needs to earn a profit. The vague demand by officialdom to recycle more often does not coincide with the technical or economic ability to make it happen. As a result, there is still a need for an explanation of how recycling occurs and why things happen as they do. Much of this second edition is about the strategy and logic behind the choice of recycling technology as well as the different toys available to the industry.

Several changes have affected the aluminum recycling industry since the first edition of this book. Among them are the following:

- The impact of Chinese demand for aluminum on global trade in scrap and the recycling industry in the rest of the world. As scrap has gravitated toward China, recyclers have closed their doors elsewhere; this has led to trade restrictions and outright export bans in some cases. Chapter 11 looks more closely at international scrap flows and politics.
- The search for a reliable way to hedge aluminum scrap purchases. This has led, in part, to the LME North American Special Aluminum Alloy Contract, which has been an issue ever since. Chapter 4 has been largely rewritten to include material on scrap purchasing and use.
- The increasing use of aluminum in the transportation industry, particularly cars and light trucks. This will change the nature of scrap supply and focus efforts on ways to sort and make optimal use of automotive scrap. Chapters 5 and 6 feature more thorough coverage of end-of-life vehicle (ELV) processing.
- The increasing reliance on old scrap as a source of recycled aluminum. Producing quality metal from less reliable input is a challenge the industry is increasingly facing. Several chapters in this edition reflect this concern.

Several of the Sherpas who helped guide me through the first edition of *Aluminum Recycling* were willing to help with the second edition. In particular, I would like to thank Ray Peterson of Aleris International, Adam Gesing of Gesing Consultants, and Don Stewart (formerly of Alcoa) for answering numerous questions and correcting even more numerous errors. Others who suffered to make this book better include Subodh Das of Phinix LLC, David DeYoung of Alcoa, John Hryn of Argonne National Laboratory, Varužan Kevorkijan of Impol, John Newby of Fives Solios, Alan Peel of Altek, Ralf Urbach of the Ingenieurberatung für Metallurgie und Rohstoffe, and John Woehlke of Evermore Recycling. As before, any errors remaining after their help are mine, although I will try and find someone else to blame.

Finally, this book was largely completed while the author was in residence as a Leif Ericsson Fellow at NTNU/SINTEF in Trondheim, Norway. My thanks go to Norsk Forskningsrådet for their support, and to Anne Kvithyld, Thorvald Abel Engh, Sarina Bao, Martin Syvertsen, and the rest of the staff for their hospitality. *Tusen takk!*

Author

Mark E. Schlesinger is a professor of metallurgical engineering at the Missouri University of Science and Technology (formerly the University of Missouri–Rolla) in Rolla, Missouri. He is a former Fulbright Scholar (Sweden) and Leif Ericsson Fellow (Norway) and dreams of someday becoming the world's shortest Viking. His teaching and research activities are centered around the fundamentals and practice of metals production. These include interests in pyrometallurgical processing, thermochemistry, phase equilibria, and of course recycling. In addition to this book and the first edition of *Aluminum Recycling*, he is also a coauthor of *Extractive Metallurgy of Copper* (fourth and fifth editions), including the material on production of secondary copper. His favorite scrap grade is Zorba.

1 Introduction

The industrial-scale production and use of aluminum metal is less than 150 years old. Yet in that time, the industry has grown until it is second only to the iron and steel industry among metal producers. The growth in aluminum usage was particularly rapid in the years following World War II, and every sector of the industry can point to products that were never produced from aluminum a generation ago but are now primarily manufactured from an aluminum alloy. Beverage cans, sports equipment, electrical buswork, window frames—all are now produced from aluminum, along with thousands of other products.

Books on the production of aluminum metal have previously focused on its recovery from naturally occurring raw materials. The principal natural ore for aluminum is bauxite, a mineral consisting primarily of hydrated aluminum oxides. Aluminum is recovered from bauxite by a selective leaching sequence known as the *Bayer process* (Damgaard et al., 2009; OECD, 2010), which dissolves most of the aluminum while leaving impurities behind. The aluminum is recovered from the leach solution by precipitating it as aluminum hydroxide. The hydroxide is then dried and calcined to generate purified alumina. The calcined alumina is in turn fed to electrolytic cells containing a molten salt electrolyte based on *cryolite* (Na_3AlF_6). The alumina dissolves in the cryolite and is electrolyzed to generate molten aluminum metal and carbon dioxide gas. This process has been the sole approach for producing *primary* aluminum metal since the late 1800s and will likely continue in this role for decades to come.

However, in the last 50 years, an increasingly large fraction of the world's aluminum supply has come from a different *ore* source. This ore is the aluminum scrap recovered from industrial waste and discarded postconsumer items. The treatment of this scrap to produce new aluminum metal and alloys is known as *recycling*, and metal produced this way is frequently termed *secondary*. Figure 1.1 shows the fraction of world aluminum production from primary and secondary sources (OECD, 2010); about one-third of the aluminum produced in the world is now obtained from secondary sources, and in some countries the percentage is much higher. As a result, the extraction of aluminum by recycling is now a topic worth describing.

The processes used for recycling aluminum scrap are very much different from those used to produce primary metal but in many ways follow the same general sequence. This sequence begins with *mining* the ore, followed by mineral processing and thermal pretreatment, and then a melting step. The metal is then refined, cast into ingots or billets or rolled into sheets, and sent to customers. Aluminum recyclers also face similar challenges to the producers of primary aluminum: the need to produce a consistent alloy with the required chemistry, minimize energy usage, reduce the amount of waste generated, and manufacture the highest-quality product at the lowest possible cost from raw materials of uncertain chemistry and condition.

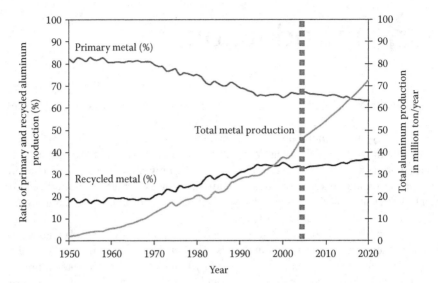

FIGURE 1.1   Total aluminum production and percent from primary and secondary sources. (From Organization for Economic Cooperation and Development [OECD], Materials case study 2: Aluminium, in *OECD Global Forum on Environment*, October 25, 2010, http://www.oecd.org/environment/resourceproductivityandwaste/46194971.pdf.)

This book will lead readers through the sequence used for recycling scrap aluminum. It will start with a description of the *minerals* (aluminum alloys) that are contained in the ore body and describe the various locations where aluminum scrap is found. It will then describe the practices used to separate scrap aluminum from the other materials with which it is mixed and the means for purifying it of coatings and other impurities. Subsequent chapters will describe the furnaces used for remelting the recovered scrap and the refining techniques used for improving its purity and quality. A final chapter will consider the unique environmental and safety challenges that recycling operations face and how these challenges are addressed.

BRIEF HISTORY OF ALUMINUM RECYCLING

As will be seen, the successful recycling of aluminum depends on several factors:

- A plentiful and recurring supply of the metal, concentrated sufficiently in one area to justify the cost of collecting it
- An infrastructure for collecting the scrap metal, removing impurities, and delivering it to a recycling facility
- A method for recycling the metal that is economically competitive with the production of the metal from natural ores
- A market for the recycled metal, should its composition or quality differ from that of primary metal

The large-scale production of aluminum metal did not begin until the 1890s, with the advent of the Hall–Heroult electrolytic process for recovery of aluminum metal

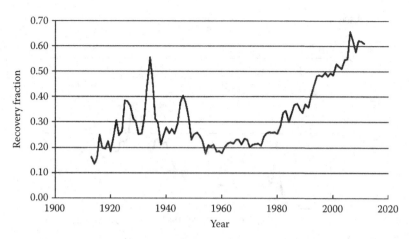

FIGURE 1.2 Fraction of aluminum consumption in the United States obtained from recycled material from 1913 through 2011. (From Buckingham, D.A. et al., *Aluminum Statistics*, U.S. Geological Survey, October 3, 2012, minerals.usgs.gov/ds/2005/140/ds140-alumi.pdf.)

from a molten salt bath. Because of this, the beginnings of aluminum recycling did not occur until the early 1900s (Aluminum Association, 1998; Hollowell, 1939).

Figure 1.2 shows the fraction of aluminum consumption in the United States (the source of the most complete statistics) obtained from recycled material from 1913 through 2011 (Buckingham et al., 2012). The changing percentages show the impact of several factors on the incentive to process scrap instead of producing primary metal:

- *Sudden increases in demand*: The importance of scrap as an aluminum supply jumped during both World Wars (Anderson, 1931; Morrison, 2005; Smith, 1946), as demand for aluminum increased faster than new primary smelters could be built. As primary smelting capacity caught up with demand, the fraction of aluminum supply obtained from secondary sources declined. Figure 1.3 shows the ratio of *old* (postconsumer) to *new* (prompt industrial) scrap used in the United States over time (Buckingham et al., 2012). The ratio is higher in the 1940s than the 1950s, due to war-time scrap drives.
- *The cost of producing primary aluminum*: Scrap became a less important source of aluminum metal in the United States during the late 1930s and 1950s. The reason was the construction of dams in the Columbia and Tennessee River valleys, which generated large amounts of low-cost hydroelectric power. This in turn encouraged the construction of new primary smelting capacity, which reduced the demand for scrap (Gitlitz, 2003). In recent years, increases in the cost of power have led to the closing of primary smelters in the United States and Europe (Home, 2009; Walker, 2012), making secondary aluminum production more significant. Figure 1.4 shows the amount of scrap used by different consumers in the United States over time (Chen and Graedel, 2012). Secondary smelters, which convert scrap into alloys suitable for casting, are a steady customer. Integrated producers, which are primary smelters and fabricators that

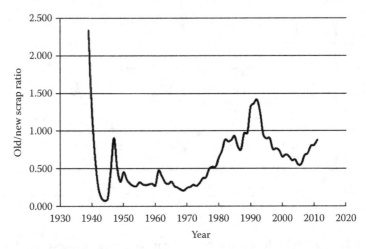

FIGURE 1.3 Ratio of old to new aluminum scrap consumption in the United States, 1939–2011. (From Buckingham, D.A. et al., *Aluminum Statistics*, U.S. Geological Survey, October 3, 2012, minerals.usgs.gov/ds/2005/140/ds140-alumi.pdf.)

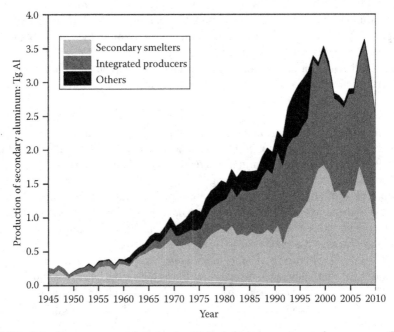

FIGURE 1.4 Scrap consumption in the United States per type of consumer. (From Chen, W.-Q. and Graedel, T.E., *Ecol. Econ.*, 81, 92, 2012. With permission.)

produce wrought alloy from primary metal or remelted low-alloy scrap, were not significant users of scrap until the 1950s. This is because in the 1950s and 1960s, the availability of cheap primary aluminum made scrap acquisition uneconomic. As primary metal production costs rose in the 1970s and 1980s, scrap began to look more attractive to integrated producers, and the fraction of

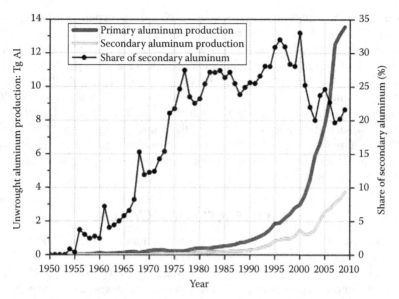

FIGURE 1.5 Historical production of aluminum in China from primary and secondary sources. (From Chen, W.-Q. and Shi, L., *Resources, Conserv. Recycl.*, 65, 18, 2012. With permission.)

used scrap recycled in the primary industry rose as well. Integrated producers currently use more scrap than primary metal (Choate and Green, 2007). Figure 1.5 illustrates aluminum production in China (Chen and Shi, 2012) and the share from secondary sources. The increasing percentage of secondary aluminum production through 2000 is the result of increasing availability of scrap as aluminum products reached the end of their useful life. The relative decline after 2000 reflects a dramatic increase in primary metal production, as the development of domestic bauxite reserves lowered the cost of production.

- *The influence of government*: Starting in the late 1960s, governments at all levels became increasingly concerned about the amount of waste discarded by industrial societies. The response was a series of laws and regulatory initiatives designed to minimize waste disposal and encourage recycling. The increase in the old/new scrap ratio from the 1970s seen in Figure 1.3 resulted largely from efforts to collect and recycle aluminum used beverage containers (UBCs). These efforts were spurred by the aluminum industry, which was attempting to avoid restrictions and mandatory deposits on the cans.
- *Improved quality*: In the early years of the industry, the secondary aluminum produced by recycling facilities was considered of inferior quality to primary metal (Anderson, 1931; Smith, 1946). This was due in part to poor chemistry control, which resulted in uncertain composition (Mahfoud and Emadi, 2010). It was also due to poor removal of dross and slags, which led to metal with too many inclusions (Dart, 2008). As tighter composition control was enforced and improved refining technology eliminated more impurities and inclusions, this stigma began to disappear.

Aluminum recycling was at first conducted *directly*, with purchased scrap being consumed by foundries without prior treatment (Aluminum Association, 1998). However, as scrap began to travel over greater distances and the reliability of its composition improved, it became desirable to produce a scrap product of greater consistency. As a result, secondary smelters began to appear after 1910 (Hollowell, 1939). These smelters produce a secondary ingot from mixed loads of scrap, with a composition matching one of several standardized grades. Table 1.1 lists the compositions of secondary alloys specified in the United Kingdom in 1946 (Smith, 1946). The alloys correspond to the primary alloys popular at the time and were used mostly in foundries. The alloys currently produced by secondary smelters will be described in Chapter 4.

ADVANTAGES (AND CHALLENGES) OF RECYCLING

There are several advantages to society when aluminum is produced by recycling rather than by primary production from bauxite ores:

Energy savings: Figure 1.6, derived from data quoted by Green (2007), shows the typical energy requirement in 1998 to produce one metric ton of primary and secondary aluminum. The high energy use (186,262 MJ/mt) for primary production is largely due to the molten salt electrolysis of alumina. Fossil fuel is required to produce the carbon electrodes and remelt the ingots produced by electrolysis, and large inputs of electrical energy are required to overcome the resistance of the electrolyte and break down the dissolved alumina in the bath. Fossil fuel use in the Bayer process (feedstock) is also a significant fraction of the total.

Direct energy usage in the production of secondary (recycled) aluminum is much smaller (11,690 MJ/mt). The largest energy user is the melting step, which is usually performed using fossil fuel. The direct energy use is reduced by 93% from that required to produce primary aluminum.

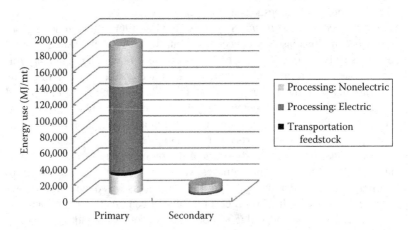

FIGURE 1.6 *Typical* energy use for primary and secondary aluminum production. (After Green, J.A.S., *Aluminum Recycling and Processing for Energy Conservation and Sustainability*, ASM International, Inc., Materials Park, OH, 2007, p. 35.)

TABLE 1.1

Chemical Compositions of Secondary Aluminum Alloys

Alloy Designation	Chemical Composition (%; Maximum Unless Otherwise Indicated)										
	Cu	Mg	Si	Fe	Mn	Ni	Zn	Pb	Sn	Ti	Other
LAC 113B	2.5–4.5	0.1	1.3	1.0	0.5	0.5	9.0–13.0	—	—	—	Pb+Ni+Sn+Mn < 1.0
ALAR 505-Z6	4.0–6.0	0.15	2.5	1.0	—	—	5.0–7.0	Pb+Sn< 0.5		—	Pb+Ni+Sn+Mn <1.0
L33	0.1	—	10.0–13.0	0.6	0.5	0.1	0.1	0.1	0.04	0.2	Modifying agents < 0.3
ALAR 00-12	0.4	0.15	10.0–13.0	0.7	0.5	0.1	0.2	0.1	0.05	0.2	Cu+Zn < 0.5Pb+ Sn < 0.1
ALAR 00-5	0.1	0.1	4.5–6.0	0.8	0.3	0.1	0.1	0.1	0.05	0.2	Pb+Sn < 0.1
LAC 112A	0.75–2.5	0.3	9.0–11.5	1.0	0.5	1.5	1.2	—	—	—	Cu+Ni < 3.0 Others < 0.5
DTD 428	6.0–8.0	0.1	2.0–4.0	1.0	—	—	2.0–4.0	—	—	—	Pb+Ni+Sn+Mn < 1.0
DTD 424	2.0–4.0	0.15	3.0–6.0	0.8	0.3–0.7	0.35	0.2	0.05	—	0.2	Fe+Mn < 1.3
LAC 10	9.0–10.5	0.15–0.35	0.6	0.3–1.0	0.6	0.5	0.1	0.1	0.1	—	Fe+Mn < 1.4 Pb+Sn+Zn<0.2
L24 "Y"	3.5–4.5	1.2–1.7	0.6	0.6	—	1.8–2.3	0.1	0.05	0.04	0.2	Si+Fe < 1.0 Sn+Zn < 0.1

Source: Smith, F.H., *Metallurgia*, 33, 207, 1946.

Reduced waste disposal: Primary aluminum production generates solid waste at every step in the process. The most significant of these are mine wastes, the red mud residue created during alumina purification, and spent pot liner from the electrolytic cells (Luo and Soria, 2008; OECD, 2010). While aluminum recycling generates solid wastes as well (primarily the dross and salt slag created during remelting), the volumes are much smaller. Ding et al. (2012) claim that the mass of solid waste generated per ton of recycled aluminum (423 kg) is 85% lower than that for primary metal (3066 kg). Lave et al. (1999) estimate that recycling aluminum reduces hazardous waste generation by over 100 kg/ton of metal produced.

Reduced emissions: Primary aluminum production generates both hazardous (fluorides, sulfur dioxide) and nonhazardous (carbon dioxide) emissions. While aluminum recycling presents its own air quality challenges, the numbers are again much reduced. Table 1.2 compares emissions from primary aluminum with that of secondary (Ding et al., 2012). The amount of CO_2 emitted per ton of metal is reduced 95% by recycling, and CO generation is reduced by more than 99%; fluoride and polycyclic aromatic hydrocarbon (PAH) emissions are almost entirely eliminated. Figure 1.7 compares several environmental indicators for the two types of aluminum production; the advantage of recycling is obvious.

Reduced capital cost: Primary aluminum production requires a mining operation, a Bayer process plant to produce purified alumina, and an electrolytic pot line to extract aluminum metal from the alumina. The capital equipment used for recycling is less complex and thus less expensive. Estimates suggest that producing aluminum by recycling rather than by primary methods reduces capital costs per ton by 90% (Choate and Green, 2007).

Challenges to the industry include the following:

Ensuring adequate supply: Scrap is a difficult sort of *ore body* to utilize, since it tends to be spread out over the landscape rather than concentrated in a single

TABLE 1.2

Air Emissions from Primary and Secondary Aluminum Production

Chemical	Primary Emission (kg/mt)	Secondary Emission (kg/mt)	Percent Reduction
CO_2	15,300	702	95.4
CO	519	1.21	99.8
SO_2	53.5	1.2	97.8
NO_x	40.4	1.79	95.6
CH_4	34	0.614	98.2
CF_4	0.0858	0	100
C_2F_6	0.0104	0	100
HF	8.01	0	100
PAH	0.0266	0	100

Source: After Ding, N. et al., *Procedia Eng.*, 27, 465, 2012.

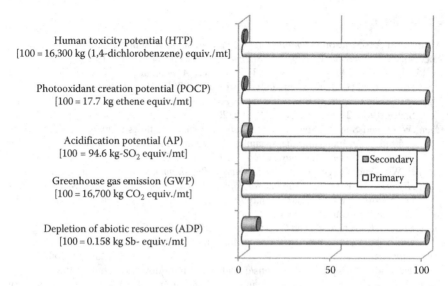

Human toxicity potential (HTP)
[100 = 16,300 kg (1,4-dichlorobenzene) equiv./mt]

Photooxidant creation potential (POCP)
[100 = 17.7 kg ethene equiv./mt]

Acidification potential (AP)
[100 = 94.6 kg-SO_2 equiv./mt]

Greenhouse gas emission (GWP)
[100 = 16,700 kg CO_2 equiv./mt]

Depletion of abiotic resources (ADP)
[100 = 0.158 kg Sb- equiv./mt]

□ Secondary
□ Primary

0   50   100

FIGURE 1.7 Environmental effect of primary and secondary aluminum production. (After Ding, N. et al., *Procedia Eng.*, 27, 465, 2012.)

location. As a result, its collection can be a problem. This is particularly true for old scrap, as will be seen in Chapter 3. The cost of collecting and processing old scrap can make this type of recycling uneconomic.

Emission control: While gaseous emissions from scrap aluminum remelting are much lower than for primary production, secondary aluminum smelters and remelters are more likely to be located in urban areas (Anderson, 1931), making them subject to tighter emissions standards (Martchek, 1997). Furthermore, the potential emissions from secondary producers are much different than those of primary producers. The processing of greasy and oily scrap, along with scrap coated with organic resins or lacquers, generates off-gases with organic compounds that must be eliminated.

Treatment and disposal of byproducts: Aluminum recycling generates solid waste products not produced in primary smelting. The most significant of these is *salt cake*, the residue of smelting under a salt flux (Aluminum Association, 1998). Another is the *shredder fluff* produced when junk automobiles are shredded to separate the recyclable metals they contain from each other. Recyclers are working on new ways to treat these solid wastes and to eliminate them if possible.

RECOMMENDED READING

Martchek, K.J., Life cycle benefits, challenges, and the potential of recycled aluminum, in *Proceedings of the Air & Waste Management Association's 90th Annual Meeting & Exhibition*, Toronto, Ontario, Canada., Paper 97-RP124B.01, 1997.

OECD (Organization for Economic Cooperation and Development), Materials case study 2: Aluminium, in OECD *Global Forum on Environment*, October 25, 2010. http://www.oecd.org/environment/resourceproductivityandwaste/46194971.pdf.

REFERENCES

Aluminum Association, *Aluminum Recycling Casebook*, Washington, DC, 1998.

Anderson, R.J., *Secondary Aluminum*, Sherwood Press, Cleveland, OH, 1931.

Buckingham, D.A., Plunkert, P.A., and Bray, E.L., *Aluminum Statistics*, U.S. Geological Survey, October 3, 2012. minerals.usgs.gov/ds/2005/140/ds140-alumi.pdf.

Chen, W.-Q. and Graedel, T.E., Dynamic analysis of aluminum stocks and flows in the United States: 1900–2009, *Ecol. Econ.*, 81, 92, 2012.

Chen, W.-Q. and Shi, L., Analysis of aluminum stocks and flows in mainland China from 1950 to 2009: Exploring the dynamics driving the rapid increase in China's aluminum production, *Resources, Conserv. Recycl.*, 65, 18, 2012.

Choate, W.T. and Green, J.A.S., *U.S. Energy Requirements for Aluminum Production: Historical Perspective, Theoretical Limits, and Current Practices*, U.S. Department of Energy, February 2007. http://www1.eere.energy.gov/manufacturing/resources/aluminum/pdfs/al_theoretical.pdf.

Damgaard, A., Larsen, A.W., and Christensen, T.H., Recycling of metals: Accounting of greenhouse gases and global warming contributions, *Waste Manage. Res.*, 27, 773, 2009.

Dart, *Aluminum Alloy: 355, 356, 357; How Important Is It Really?*, May 20, 2008. www.dartheads.com/wp-content/plugins/download-monitor/download.php?id=198.

Ding, N., Gao, F., Wang, Z., Gong, X., and Nie, Z., Environment impact analysis of primary aluminum and recycled aluminum, *Procedia Eng.*, 27, 465, 2012.

Gitlitz, J., The role of the consumer in reducing primary aluminum demand, Container Recycling Institute, Culver City, CA, 2003. http://www.container-recycling.org/assets/pdfs/aluminum/Aluminum-RoleofConsumer.pdf.

Green, J.A.S., *Aluminum Recycling and Processing for Energy Conservation and Sustainability*, ASM International, Inc., Materials Park, OH, 2007, p. 35.

Hollowell, R.D.T., Secondary aluminium and its alloys, *Met. Ind.*, 54, 387, 1939.

Home, A., Metals insider—European aluminium smelters an endangered species, December 23, 2009. http://in.reuters.com/article/2009/12/23/metals-insider-idINLDE5BM0UP20091223.

Lave, L.B., Hendrickson, C.T., Conway-Schempf, N.M., and McMichael, F.C., Municipal solid waste recycling issues, *J. Environ. Eng.*, 125, 944, 1999.

Luo, Z. and Soria, A., Prospective study of the world aluminium industry, *JRC Sci. Tech. Rpt.*, EUR 22951, February 7, 2008. http://ftp.jrc.es/EURdoc/JRC40221.pdf.

Mahfoud, M. and Emadi, D., Aluminum recycling—Challenges and opportunities, *Adv. Mater. Res.*, 83–86, 571, 2010.

Martchek, K.J., Life cycle benefits, challenges, and the potential of recycled aluminum, in *Proceedings of the Air & Waste Management Association's 90th Annual Meeting & Exhibition*, Toronto,Ontario,Canada., Paper 97-RP124B.01, 1997.

Morrison, J., European aluminium recycling under threat?, *Alum. Int. Today*, 17(1), 17, January 2005.

OECD (Organization for Economic Cooperation and Development), Materials case study 2: Aluminium, in *OECD Global Forum on Environment*, October 25, 2010. http://www.oecd.org/environment/resourceproductivityandwaste/46194971.pdf.

Smith, F.H., Secondary aluminium, *Metallurgia*, 33, 207, 1946.

Walker, S., Aluminum: Smelters struggle as prices fall, *WOMP*, 11, 2012. http://www.womp-int.com/story/2012vol03/story026.htm.

2 Ore Body

In traditional extractive metallurgy, the raw material used to produce a metal is an ore mined from the earth. The ore contains *minerals,* chemical compounds that include the desired metallic element. It also contains other chemical compounds that do not include the metallic element; these are known as *gangue* minerals. The ore mineral contains *impurity* elements in addition to the metallic element. These must be chemically separated from the metal during the extraction process.

In aluminum recycling, the *ore body* consists of scrap metal found on the ground, rather than in it. However, the similarities are greater than what might appear at first. Aluminum scrap comes with gangue minerals of its own, in the form of attached items and coatings. It also includes impurity elements that can have a significant influence on the recycling process. As a result, a description of the ore body for aluminum recycling will explain why some processes and strategies have been adopted.

This chapter will introduce the families of aluminum alloys and the products made from them. It is not meant to be an extensive discussion. Other sources provide a much more complete description of aluminum metallurgy and products, and the reader is encouraged to consult these for more information. The purpose of this chapter is to describe the raw material that the recycler obtains and how it impacts the available process options.

WROUGHT ALUMINUM ALLOY SYSTEM (ALTENPOHL, 1998; KAUFMAN, 2000)

As the name suggests, wrought aluminum alloys are those that are turned into consumer products by a solid-state process—extrusion, forging, or rolling. There are over 200 such alloys. Nearly all consist of at least 90% aluminum, and most over 95%. Wrought aluminum alloys are divided into eight classes based on the choice of alloying element, according to a system developed in the United States in the 1950s. Table 2.1 lists the composition of several common aluminum alloys from the different classes.

1xxx alloys are the purest aluminum alloys, containing 99% or more aluminum. The high aluminum content ensures high electrical conductivity and corrosion resistance but results in lower strength than other alloys. 1xxx alloys are used in electrical applications, as well as for packaging foil and for chemical equipment. Because the minimum required aluminum percentage is so high, it is difficult to use scrap for producing these alloys. As a result, only primary aluminum and carefully selected 1xxx alloy scrap are used. However, 1xxx scrap can be readily used for producing alloys from any other series.

2xxx alloys contain 1.0%–6.0% copper, depending on the alloy. Other alloying elements in 2xxx alloys include iron, magnesium, manganese, and/or silicon. These are among the highest-strength aluminum alloys and are also noted for

TABLE 2.1

Composition of Common Wrought Aluminum Alloys

Designation	% Si	% Fe	% Cu	% Mn	% Mg	% Cr	Other	% Zn	% Ti	Others Each	Others Total
1050	0.25	0.40	0.05	0.05	0.05	—	0.05 V	0.05	0.03	0.03	>99.5 Al
1100	0.95 Fe + Si		0.050–0.20	0.05	—	—	—	0.10	—	0.05	>99.0 Al
1175	0.15 Fe + Si		0.10	0.02	0.02	—	0.03 Ga; 0.05 V	0.04	0.02	0.02	>99.75 Al
1350	0.10	0.40	0.05	0.01	—	0.01	—	0.05	—	0.03	0.10
2007	0.80	0.80	3.3–4.6	0.50–1.0	0.40–1.8	0.10	0.20 Ni; 0.20 Bi; 0.80–1.5 Pb, 0.20 Sn	0.80	0.20	0.10	0.30
2011	0.40	0.70	5.0–6.0	—	—	—	0.20–0.60 Bi; 0.20–0.60 Pb	0.30	—	0.05	0.15
2017	0.20–0.80	0.70	3.5–4.5	0.40–1.0	0.40–0.80	0.10	—	0.25	0.15	0.05	0.15
2024	0.50	0.50	3.8–4.9	0.30–0.90	1.2–1.8	0.10	—	0.25	0.15	0.05	0.15
2038	0.50–1.3	0.60	0.80–1.8	0.10–0.40	0.40–1.0	0.20	—	0.50	0.15	0.05	0.15
2048	0.15	0.20	2.8–3.8	0.20–0.60	1.2–1.8	—	—	0.25	0.10	0.05	0.15
2090	0.10	0.12	2.4–3.0	0.05	0.25	0.05	1.9–2.6 Li	0.10	0.08–0.15 Zr	0.05	0.15
2091	0.20	0.30	1.8–2.5	0.10	1.1–1.9	0.10	1.7–2.3 Li	0.25	0.04–0.16 Zr	0.05	0.15
2117	0.20–0.80	0.7	3.5–4.5	0.40–1.0	0.40–1.0	0.10	—	0.25	0.25 Zr + Ti	0.05	0.15
2124	020	0.30	3.8–4.9	0.30–0.90	1.2–1.8	0.10	—	0.25	0.15	0.05	0.15
2219	0.20	0.30	5.8–6.8	0.20–0.40	0.02	—	0.05–0.15 V; 0.10–0.25 Zr	0.10	0.02–0.10	0.05	0.15
2224	0.12	0.15	3.8–4.4	0.30–0.90	1.2–1.8	0.10	—	0.25	0.15	0.05	0.15

2324	0.10	0.12	3.8–4.4	0.30–0.90	1.2–1.8	0.10	—	0.25	0.15	0.05	0.15
2618	0.10–0.25	0.90–1.3	1.9–2.7	—	1.3–1.8	—	0.90–1.2 Ni	0.10	0.04–0.10	0.05	0.15
3003	0.60	0.7	0.05–0.20	1.0–1.5	—	—	—	0.10	—	0.05	0.15
3004	0.30	0.70	0.25	1.0–1.15	0.80–1.3	—	—	0.25	—	0.05	0.15
3005	0.60	0.70	0.30	1.0–1.5	0.20–0.60	0.10	—	0.25	0.10	0.05	0.15
3103	0.50	0.70	0.10	0.90–1.5	0.30	0.10	—	0.20	0.10 Ti + Zr	0.05	0.15
3104	0.60	0.80	0.05–0.25	0.80–1.4	0.80–1.3	—	0.05 Ga; 0.05 V	0.25	0.10	0.05	0.15
4032	11.0–13.5	1.0	0.50–1.3	—	0.80–1.3	0.10	0.50–1.3 Ni	0.25	—	0.05	0.15
4043	4.5–6.0	0.8	0.30	0.05	0.05	—	—	0.10	0.20	0.05	0.15
5005	0.30	0.70	0.20	0.20	0.50–1.1	0.10	—	0.25	—	0.05	0.15
5017	0.40	0.70	0.18–0.28	0.60–0.80	1.9–2.2	—	—	—	0.09	0.05	0.15
5042	0.20	0.35	0.15	0.20–0.50	3.0–4.0	0.10	—	0.25	0.10	0.05	0.15
5052	0.25	0.40	0.10	0.10	2.2–2.8	0.15–0.35	—	0.10	—	0.05	0.15
5082	0.20	0.35	0.15	0.15	4.0–5.0	0.15	—	0.25	0.15	0.05	0.15
5083	0.40	0.40	0.10	0.40–1.0	4.0–4.9	0.05–0.25	—	0.25	0.15	0.05	0.15
5182	0.20	0.35	0.15	0.20–0.50	4.0–5.0	0.10	—	0.25	0.10	0.05	0.15
5252	0.08	0.10	0.10	0.10	2.2–2.8	—	0.05 V	0.05	—	0.03	0.10
5356	0.25	0.40	0.10	0.05–0.20	4.5–5.5	0.05–0.20	—	0.10	0.06–0.20	0.05	0.15
5454	0.25	0.40	0.10	0.50–1.0	2.4–3.0	0.05–0.20	—	0.25	0.20	0.05	0.15
5457	0.08	0.10	0.20	0.15–0.45	0.80–1.2	—	0.05 V	0.05	—	0.03	0.10
5657	0.80	0.10	0.10	0.03	0.60–1.0	—	0.03 Ga; 0.05 V	0.05	—	0.02	0.05
5754	0.40	0.40	0.10	0.50	2.6–3.6	0.30	—	0.20	0.15	0.05	0.15
6005	0.60–0.90	0.35	0.10	0.10	0.40–0.60	0.10	—	0.10	0.10	0.05	0.15
6016	1.0–1.5	0.50	0.20	0.20	0.25–0.60	0.10	—	0.20	0.15	0.05	0.15
6060	0.30–0.60	0.10–0.30	0.10	0.10	0.35–0.60	0.50	—	0.15	0.10	0.05	0.15
6061	0.40–0.80	0.70	0.15–0.40	0.15	0.80–1.2	0.04–0.35	—	0.25	0.15	0.05	0.15
6063	0.20–0.60	0.35	0.10	0.10	0.45–0.90	0.10	—	0.10	0.10	0.05	0.15

(continued)

TABLE 2.1 (continued)
Composition of Common Wrought Aluminum Alloys

Designation	% Si	% Fe	% Cu	% Mn	% Mg	% Cr	Other	% Zn	% Ti	Others Each	Others Total
6082	0.70–1.3	0.50	0.10	0.40–1.0	0.60–1.2	0.25	—	0.20	0.10	0.05	0.15
6111	0.70–1.1	0.40	0.50–0.90	0.15–0.45	0.50–1.0	0.10	—	0.15	0.10	0.05	0.15
7005	0.35	0.40	0.10	0.20–0.70	1.0–1.8	0.60–1.2	0.08–0.2 Zr	4.0–5.0	0.01–0.06	0.05	0.15
7020	0.35	0.40	0.20	0.05–0.50	1.0–1.4	0.10–0.35	0.08–0.20 Zr	4.0–5.0	—	0.05	0.15
7075	0.40	0.50	1.2–2.0	0.30	2.1–2.9	0.18–0.28	—	5.1–6.1	0.20	0.05	0.15
7150	0.12	0.15	1.9–2.5	0.10	2.0–2.7	0.04	0.08–0.15 Zr	5.9–6.9	0.06	0.05	0.15
7175	0.15	0.20	1.2–2.0	0.10	2.1–2.9	0.18–0.28	—	5.1–6.1	0.10	0.05	0.15
7178	0.40	0.50	1.6–2.4	0.30	2.4–3.1	0.18–0.28	—	6.3–7.3	0.20	0.05	0.15
7475	0.10	0.12	1.2–1.9	0.06	1.9–2.6	0.18–0.25	—	5.2–6.2	0.06	0.05	0.15
8011	0.50–0.90	0.60–1.0	0.10	0.20	0.05	0.05	—	0.10	0.08	0.05	0.15
8017	0.10	0.55–0.80	0.10–0.20	—	0.01–0.05	—	0.04 B, 0.0003 Li	0.05	0.10	0.05	0.15
8090	0.20	0.30	1.0–1.6	0.10	0.60–1.3	0.10	2.2–2.7 Li	0.25	0.04–0.16 Zr	0.05	0.15

Source: Cayless, R.B.C., Alloy and temper designation systems for aluminum and aluminum alloys, in *Metals Handbook, Vol. 2: Properties and Selection: Nonferrous Alloys and Special-Purpose Materials*, 10th edn., ASM International, Materials Park, OH, 1990. With permission.

their toughness. They are widely used in aircraft and for fasteners as well. 2xxx alloys are not very corrosion resistant and so are typically painted or clad before being put in service. Many aluminum alloys outside the 2xxx series are limited to 0.30% copper or less (Bijlhouwer, 2009), and so 2xxx scrap cannot be used to produce wrought alloys from most other classes (Froelich et al., 2007).

The number of *3xxx* series alloys is limited, but their use is widespread. These alloys feature up to 1.5% manganese and usually contain 0.7%–0.8% iron. They can be strain hardened and have excellent corrosion resistance. These alloys are best known for their use in the bodies of aluminum beverage cans (Das et al., 2010). They are also widely used in cooking utensils, automobile radiators, roofing and siding, and heat exchangers. Restrictions on the use of 3xxx scrap center around the manganese and iron content, which are often unwelcome in other aluminum alloys.

The number of *4xxx* alloys is also limited, and these alloys are not as widely used as other classes. 4xxx alloys contain up to 13% silicon, which improves their wear resistance. Their primary use is in forgings such as aircraft pistons. They are also used for welding other aluminum alloys. The high silicon content makes them largely unusable for recycling into anything but casting alloys or 4xxx wrought alloys (Froelich et al., 2007).

5xxx alloys feature up to 5.5% magnesium. They are especially noted for their corrosion resistance and toughness and are easily welded. As a result, the use of 5xxx alloys is widespread. They are used in items as large as bridges and storage tanks and as small as the lids of beverage cans (Das et al., 2010). They are also increasingly popular in automotive structural applications, notably body panels. Magnesium is the only alloying element that can be easily refined from molten aluminum. As a result, 5xxx alloys are more recyclable for wrought-alloy production than most other alloys. However, the cost of magnesium recommends that these alloys be recycled back into 5xxx compositions if possible.

The *6xxx* alloys also feature up to 1.5% magnesium, along with silicon levels up to 1.8%. The presence of these two elements makes them heat treatable, and the low overall alloy content gives them extrudability. Their best-known uses are in architecture, but they are also used in automotive applications, in welded structural applications, and as high-strength conductor wires. As with 4xxx alloys, the primary limitation on their recyclability is the silicon content.

7xxx series alloys are the most heavily alloyed of the wrought aluminum alloys. They feature 1.5%–10% zinc, depending on the alloy. 7xxx alloys also have up to 3% magnesium, and some feature up to 2.6% copper. Heat-treated 7xxx alloys have the highest strengths of the common wrought aluminum alloys and impressive toughness levels. As with the 2xxx alloys, their lower corrosion resistance usually results in them being painted or coated when put in service. Historically, their greatest use has been in the aircraft industry, but they have also found use in transport applications. Their zinc and copper content make 7xxx alloys difficult to recycle into anything except 7xxx alloy or some casting alloys (Froelich et al., 2007).

8xxx alloys feature alloying elements not used in the other series. Boron, iron, lithium, nickel, tin, and vanadium are some of the elements used in 8xxx alloys. Because this is a catch-all category, 8xxx alloys cannot be characterized in terms of property or use. One notable trend in aluminum alloy development reflected in this series is the increasing number of lithium-containing aluminum alloys. These alloys have an especially high modulus of elasticity and are increasingly used in aircraft construction to save weight without sacrificing stiffness. Lithium is also added to some 2xxx alloys. The recycling of lithium-containing alloys is a particular challenge, since the value of the lithium contained in these parts recommends closed-loop recycling into the same alloy if at all possible.

As previously suggested, the difficulty in recycling wrought aluminum alloys is the problem of *tolerance,* the ability of an alloy to absorb elements not normally present in its composition (Kevorkijan, 2010). To provide an example, the wrought aluminum fraction of an automobile will be considered as a potential charge to a recycling furnace. A simplified breakdown for the wrought aluminum might consist of the following (Zapp et al., 2002):

- 35% alloy 6060 (0.45% Si, 0.2% Fe, 0.5% Mg)
- 11% alloy 6082 (1.0% Si, 0.3% Fe, 0.7% Mn, 0.9% Mg)
- 10% alloy 3003 (0.5% Si, 0.5% Fe, 1.3% Mn)
- 9% alloy 5182 (0.1% Si, 0.2% Fe, 0.4% Mn, 4.5% Mg)
- 14% alloy 5754 (0.3% Si, 0.2% Fe, 0.4% Mn, 3.2% Mg)
- 15% alloy 6016 (1.25% Si, 0.3% Fe, 0.5% Mg)
- 6% alloy 7020 (0.2% Si, 0.3% Fe, 0.3% Mn, 1.2% Mg, 4.5% Zn)

Melted down, this mixture would produce an alloy analyzing 0.57% Si, 0.26% Fe, 0.32% Mn, 1.27% Mg, and 0.27% Zn. The zinc content is higher than the allowed maximum for many common alloys outside the 7xxx series. However, the silicon content is higher than allowed in 7xxx compositions. As a result, the wrought scrap cannot be used to produce new wrought alloy unless one or more of the impurities are diluted by the addition of primary metal. If cast scrap is included in the mix, the problem becomes even worse. Table 2.2, adapted from the work of Froelich et al. (2007), shows the compositional compatibility of different classes of cast and wrought aluminum alloy. The number of − markers indicating incompatibility illustrates the tolerance problem. The development of *recycle-friendly* alloys with impurity limits that can tolerate higher scrap inputs from other systems has become a research priority within the aluminum industry (Das, 2007; Gaustad et al., 2010).

CAST ALUMINUM ALLOY SYSTEM (ALTENPOHL, 1998; KAUFMAN, 2000)

There are now over 200 cast aluminum alloys in addition to the wrought compositions. Cast alloys tend to have higher alloy content than wrought alloys. This makes them difficult to recycle into anything other than cast alloy, since the removal of most alloying elements from molten aluminum is impractical. As is the case for wrought alloys, cast alloys are divided into classes based on the alloying elements used, using a system devised in the 1950s. The numbering of these alloys is meant to correspond roughly to similar classes of wrought alloys. Table 2.3 lists the compositions of the most commonly used cast alloys.

1xx.x cast alloys are quite rare in service but have aluminum contents of 99% or higher, similar to 1xxx wrought alloys. Generally, they are produced only as ingot, for remelting and alloying to a different composition. Their use as consumer

TABLE 2.2

Compatibility of Aluminum Alloy Classes with Each Other for Recycling Tolerance

Alloy Class	300	200	300[a]	300[b]	300[c]	500	400	1000	3000	5000	6000	7000
300	+											
200	−	+										
300[a]	+	0	+									
300[b]	+	−	0	+								
300[c]	−	−	−	0	+							
500	0	−	−	−	−	+						
400	−	−	−	0	+	0	+					
1000	−	−	−	−	+	−	−	+				
3000	−	−	−	−	−	0	−	−	+			
5000	−	−	−	−	−	+	−	−	−	+		
6000	−	−	−	0	0	−	0	−	−	−	+	
7000	−	−	−	−	−	−	−	−	−	−	−	+

Source: Adapted from Froelich, D. et al., *Miner. Eng.,* 20, 891, 2007.

Note: +, Compatibility; 0, some compatibility; −, incompatible.

[a] +Mg, Fe/Cu/Mn.

[b] +Mg, Fe.

[c] +Mg.

TABLE 2.3

Composition of Common Cast Aluminum Alloys

Designation	% Si	% Fe	% Cu	% Mn	% Mg	% Cr	Other	% Zn	% Ti	Others Each	Others Total
201.0	0.10	0.15	4.0–5.2	0.20–0.50	0.15–0.55	—	0.24–1.0 Ag	—	0.15–0.35	0.05	0.10
203.0	0.30	0.50	4.5–5.5	0.20–0.30	0.10	—	1.3–1.7 Ni; 0.20–0.30 Sb; 0.20–.30 Co; 0.10–0.30 Zr	—	0.15–0.25	0.05	0.15
295.0	0.7–1.5	1.0	4.0–5.0	0.35	0.03	—	—	0.35	0.25	0.05	0.15
308.0	5.0–6.0	1.0	4.0–0	0.50	0.10	—	—	1.0	0.25	—	0.50
319.0	5.5–6.5	1.0	3.0–4.0	0.50	0.10	—	0.35 Ni	1.0	0.25	—	0.50
332.0	8.5–10.5	1.2	2.0–4.0	0.50	0.5–1.5	—	0.50 Ni	1.0	0.25	—	0.50
347.0	6.5–7.5	0.15	0.05	0.03	0.45–0.60	—	—	0.05	0.20	0.05	0.15
355.0	4.5–5.5	0.60	1.0–1.5	0.50	0.40–0.60	0.25	—	0.35	0.25	0.05	0.15
356.0	6.5–7.5	0.60	0.25	0.35	0.20–0.45	—	—	0.35	0.25	0.05	0.15
357.0	6.5–7.5	0.20	0.20	0.10	0.40–0.70	—	0.04–0.07 Be	0.10	0.04–0.20	0.03	0.15
A356.0	6.5–7.5	0.20	0.20	0.10	0.25–0.45	—	—	0.10	0.20	0.05	0.15
359.0	8.5–9.5	0.20	0.20	0.10	0.50–0.70	—	—	0.10	0.20	0.05	0.15
A360.0	9.0–10.0	1.3	0.60	0.35	0.40–0.60	—	0.5 Ni	0.50	—	—	0.25

361.0	9.5–10.5	1.1	0.50	0.25	0.40–0.60	0.20–0.30	0.2–0.3 Ni; 0.1 Sn	0.50	0.20	0.05	0.15
364.0	7.5–9.5	1.5	0.20	0.10	0.20–0.40	0.25–0.50	0.02–0.04 Be; 0.15 Sn; 0.15 Ni	2.15	—	0.05	0.15
380.0	7.5–9.5	2.0	3.0–4.0	0.50	0.10	—	0.5 Ni; 0.15 Sn	3.0	—	—	—
A390.0	16.0–18.0	0.50	4.0–5.0	0.10	0.45–0.65	—		0.10	0.20	0.10	0.20
A413.0	11.0–13.0	1.3	1.0	0.35	0.10	—	0.15 Sn	3.0	—	—	0.25
B413.0	11.0–13.0	0.50	0.10	0.35	0.05	—	—	0.50	—	—	0.20
443.0	4.5–6.0	0.80	0.60	0.50	0.05	0.25	—	0.15	0.25	—	0.35
512.0	1.4–2.2	0.60	0.35	0.80	3.5–4.5	0.25	—	0.50	0.25	0.05	0.35
514.0	0.30	0.30	0.10	0.10	3.6–4.5	—	—	1.4–2.2	0.20	0.05	0.15
520.0	0.25	0.30	0.25	0.15	9.5–10.6	—	—	0.15	0.25	0.05	0.15
535.0	0.15	0.15	0.05	0.10–0.25	6.2–7.5	0.25	0.003–0.007 Be; 0.005 B	0.35	0.25	0.05	0.15
705.0	0.20	0.80	0.20	0.40–0.60	1.4–1.8	0.20–0.40	—	2.7–3.3	0.25	0.05	0.15
850.0	0.70	0.70	0.70–1.3	0.10	0.10	—	0.7–1.3 Ni; 5.5–7.0 Sn	—	0.20	—	0.30

Source: Cayless, R.B.C., Alloy and temper designation systems for aluminum and aluminum alloys, in *Metals Handbook, Vol. 2: Properties and Selection: Nonferrous Alloys and Special–Purpose Materials*, 10th edn., ASM International, Materials Park, OH, 1990. With permission.

products is limited to applications requiring high electrical conductivity, such as pressure-cast integral conductor bars.

2xx.x alloys all contain 3.5%–10.7% copper, just as 2xxx wrought alloys feature copper as the primary alloying agent. 2xx.x alloys can also have significant levels of iron, magnesium, nickel, or silicon. These alloys have the highest high-temperature strength and hardness of the cast alloys and require heat treatment to prevent stress corrosion cracking. As with the 2xxx wrought alloys, coating or painting is usually required by the lower corrosion resistance of these alloys.

The *3xx.x* alloys are the most widely used of all cast aluminum alloys. They have silicon levels ranging from 4.5% to over 20% and copper levels ranging from 0.5% to 5.0%. Some 3xx.x alloys also include magnesium (0.2%–1.5%) and a few have nickel (0.5%–3.0%). The high silicon content increases fluidity, reduces cracking, and minimizes shrinkage porosity in castings. The copper and magnesium provide solid solution hardening. The nickel reduces the coefficient of thermal expansion (as does higher silicon content), which is useful in pistons and cylinders.

4xx.x alloys also feature 3.3%–13% silicon but much smaller levels of other elements. Copper, iron, or nickel are sometimes added. Their ductility and impact resistance make them useful in applications such as food-handling equipment and marine fittings.

As is the case with 5xxx wrought alloys, the specifications for *5xx.x* cast alloys include 3.5%–10.5% magnesium. Smaller levels of iron, silicon, and zinc are sometimes used. Along with 2xx.x, 7xx.x, and 8xx.x alloys, 5xx.x alloys are not as castable as high-silicon compositions, requiring some care in mold design.

The *7xx.x* cast alloy specifications require 2.7%–8.0% zinc, similar to their 7xxx wrought counterparts. Magnesium (0.5%–2.0%) is present in many of these alloys, and chromium (0.2%–0.6%) is sometimes added as well. They have better corrosion resistance than other cast alloys and develop high strength through natural aging, eliminating the need for heat treatment.

Finally, the *8xx.x* series of cast alloys employ 5.5%–7.0% tin and 0.7%–4.0% copper, which provide strength and lubricity. These alloys are used for bearing applications, such as connecting rods and crankcase bearings. 8xx.x alloys are much less used than alloys from other series.

As with the wrought alloys, the difficulty in recycling cast alloys is one of tolerance. Most alloys outside the 7xx.x series have a strict upper limit on the allowable level of zinc, while some 3xx.x alloys have low maximum allowable levels of iron. An exception to this is alloy 380.0, which has reasonably high levels of all the major alloying elements. As a result, 380.0 (7.5%–9.5% Si, 1.5% Fe, 3.0%–4.0% Cu, 0.5% Mn, 0.5% Ni, <0.1% Mg, <0.15% Sn) is often the main product of secondary smelters trying to make use of all the available cast scrap. This alloy can then be diluted by customers with 1xxx or 1xx.x scrap, or with primary metal, to produce more desirable alloys.

For example, if 100 kg of 380.0 scrap (8.5% Si, 2.0% Fe, 3.5% Cu, 0.5% Mn, 0.5% Ni, 0.1% Sn, 3.0% Zn) is diluted with 200 kg of primary metal (0.05% Si, 0.06% Fe), the result will be 300 kg of metal analyzing 2.87% Si, 0.71% Fe, 1.17% Cu, 0.17% Mn, 0.17% Ni, 0.03% Sn, and 1.00% Zn. By adding additional amounts of some elements, the melter can create casting alloy 332.0 (9.5% Si, 1.2% Fe, 3.0% Cu, 0.5 Mn, 1.0% Mg, 0.5% Ni, 1.0% Zn, 0.25% Ti), a popular alloy for automotive castings. The required level of dilution with primary aluminum is substantial.

However, this tactic allows the recycling of all the cast aluminum scrap, without the need to separate one alloy from another.

PRODUCT MIX (ALTENPOHL, 1998)

The market for aluminum products is generally separated into seven segments—building and construction, transportation, consumer durables, electrical, machinery and equipment, packaging, and other. Figure 2.1 illustrates the fraction of aluminum products falling into these applications in several countries. As can be seen, the product mix varies substantially from one country to another, which ultimately affects the nature of the scrap supply.

The properties of aluminum that make it most valuable in the *building and construction* sector are its low density, its high corrosion resistance, and the design flexibility resulting from the ease with which aluminum can be extruded. Examples of applications benefiting from these advantages include the following:

- Mobile military bridges
- Window frames
- Highway guard rails
- Loading ramps
- Greenhouses
- Building cladding and facades
- Flag poles
- Mobile and bridge cranes
- Ladders
- Automatic doors
- Road sign supports
- Cable support towers

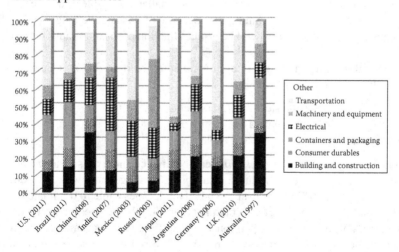

FIGURE 2.1 Aluminum use by sector in various countries. (Adapted from Menzie, W.D. et al., The global flow of aluminum from 2006 through 2025, USGS Open-File Report 2010-1256, http://pubs.usgs.gov/of/2010/1256/pdf/ofr2010-1256.pdf, 2010.)

Most of the alloys used in this sector are wrought alloys, particularly from the 5xxx and 6xxx series (Jones, 2002). The more commonly used alloys are 3103, 3105, 5005, 5083, 5085, 5754, 6005, 6060, 6061, 6063, and 6082. Much of the use of aluminum in this sector is relatively new, and the lifetime of the products in which aluminum is used is fairly lengthy—20 years is common, and some items may be in service as long as 50 years (McMillan et al., 2010; Menzie et al., 2010). As a result, the share of old scrap derived from building and construction debris has increased from almost zero in 1999 to 8% of the world supply (IAI, 2009) and will increase further.

Gangue materials and potential contaminants in aluminum scrap recovered from building and construction debris include:

- Ceramic materials, including brick, concrete, and glass
- Organic coatings, such as polyvinylidene fluoride (Jones, 2002)
- Dirt and paint
- Polymers, including plastic and rubber
- Steel pieces, including those attached to the aluminum

As Figure 2.1 points out, the fraction of aluminum used for *electrical engineering* applications in developing countries is much higher than in developed countries (Menzie et al., 2010). This is the result of continuing construction of the energy grid in developing countries, which uses large amounts of aluminum cable. However, aluminum use in wire and cable is growing in developed countries as well (Aluminum Association, 2012), due to substitution encouraged by the high price of copper (Kooiman, 2011). Alloys 1350 and 6201 have been traditionally used for wire and rod and alloy 6101 for bus bars and other large conductors; however, 8000-series alloys such as 8017 and 8176 have superior creep resistance, which makes them more suitable as building wire (Southwire, 2012). Their use complicates recycling efforts, since 6xxx and 8xxx scrap cannot be used to produce new 1xxx alloy. A small amount of 1xx.x cast alloy is also used. For smaller size cable and wire, builders continue to favor copper over aluminum. Because of this and the long service life of these products, wire and cable will be a small fraction of the aluminum scrap supply in the future.

Potential gangue impurities in aluminum scrap recovered from electrical products include the following:

- Steel supports from steel-reinforced cable
- Copper wire mixed with aluminum wire
- Plastics used to coat wire and cable

The use of aluminum in *packaging* can be divided into two subcategories: rigid containers (cans) and foil products. Aluminum beverage cans were introduced in the United States in 1959 and over the next two decades came to dominate the market for small soft drink and beer containers, driving the competing steel can out of this market. While some competition from plastic and glass remains, the average American now uses 350 aluminum cans per year. Several factors have contributed to this, including the recyclability of the can. Aluminum is the only packaging material that has significant value as a recyclable material. This generates an incentive to recover the cans, resulting in a higher recycling rate than for plastic or glass.

In Europe and Japan, the popularity of aluminum cans has not spread as quickly as in North America. Per capita use of aluminum cans in Europe and South America is about one-seventh of that in the United States (Woehlke, 2011). In Japan, the aluminum can has become predominant over the steel can for soft drink and beer containers but still faces significant competition from glass bottles and paper cartons (Inada, 2010; JAA, 2010). Aluminum is also used in food cans and aerosol containers but is not as popular in these applications (Golden, 2009).

The typical aluminum beverage can is produced from two alloys (Woehlke, 2011). The can body (83% of the total weight) is produced from alloy 3004. The lid and tab (17% of the total weight) is made of alloy 5182. This mixture presents a challenge in recycling the can, since the alloy produced by simply remelting the can has too much magnesium to be used as 3004 and too much manganese to be used as 5182. As a result, UBCs must be diluted after melting with primary metal or low-alloy scrap in order to produce more cans. The aluminum industry has made efforts to develop a single alloy that can be used for both the body and lid (Nishiyama, 2002); this would allow for truly direct recycling.

Aluminum strip and foil is also used to produce a variety of pouches, food trays, and single-serve beverage containers (Golden, 2009). These containers are often produced from rolled 1xxx alloys or 8011 (Gesing, 2001) and so ought to be easily recycled. However, packages of this sort are composite structures, with large quantities of paper and plastic, which must be separated from the thin aluminum foil before the metal can be remelted. Because of this, the cost of recycling per unit weight of metal recovered is high. This reduces the incentive to recover foil- and strip-based packaging from the waste stream and results in much lower recycling rates for this type of aluminum.

Contaminants in aluminum recovered from packaging applications include, among others, the following (Evermore Recycling, 2012):

- Paper and plastic, especially from pouches and composite boxes
- Steel cans, in recovered UBCs
- Rocks, dirt, sand, and glass
- Organic coatings and paint on can surfaces
- Batteries
- Excessive moisture

A variety of cast and wrought aluminum alloys are used to produce *consumer durables*. Examples include the following:

- 3003 (cookware)
- 3004 (lamp bases)
- 5005 (utensils, appliances)
- 5052, 6463 (appliances)
- 5056 (insect screens)
- 5457, 5657 (appliance trim)
- 6005 (TV antennas)
- 6061 (furniture, canoes)
- 6063 (furniture)

- 360.0 (cookware, instrument cases, cover plates)
- 380.0 (lawnmowers, dental equipment)
- B443.0 (waffle irons)
- 513.0 (ornamental)

However, the majority of aluminum use as consumer durables is for so-called *white goods* (Cooper, 2010). These include refrigerators, freezers, ranges, dishwashers, and clothes washers and dryers. The percentage of aluminum in white goods is small (CAMA, 2005), but the number produced each year makes them a substantial aluminum market nonetheless.

Generalizing about this category is not possible, given the diversity. However, scrap from this class of items is more likely to come with attached contaminants than other types of aluminum scrap and thus requires more processing. This kind of scrap is also less likely to be deliberately collected than others, meaning that its collection rate is usually lower.

The use of aluminum in *transport* applications has long affected recycling patterns. In fact, the first major surge in recycling rates resulted from the scrapping of aluminum parts from surplus aircraft following World War I (Morrison, 2005). More recently, the transport sector has become the largest user of aluminum, meaning that used vehicles are an increasingly substantial fraction of the available scrap supply (McMillan et al., 2010). As a result, the patterns of aluminum alloy use in transportation are especially significant.

Aluminum continues to be the primary material for aircraft construction, despite inroads made by composites. 2xxx alloys have long been heavily used in aircraft construction, in particular 2024, 2124, 2224, and 2324. Some 7xxx alloys (7055, 7075, 7150) are also used. In recent years, lithium-containing alloys (2090, 2091, 8090) have become part of the mix. Aluminum is also extensively used in passenger ships, particularly 5xxx and 6xxx alloys (JAA, 2010). However, as Figure 2.2 shows

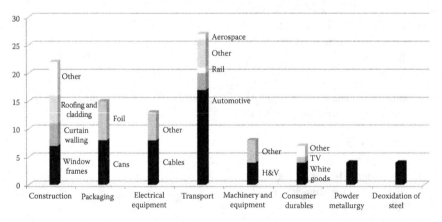

FIGURE 2.2 Aluminum use by industry sector. (Reproduced from Cooper, D.R., Steel and aluminium intensive products: Their metallic components and design requirements, http://www.lcmp.eng.cam.ac.uk/wp-content/uploads/W6-Steel-and-aluminium-products.pdf, 2010. With permission.)

(Cooper, 2010), these sectors are a very small fraction of aluminum transport use. Aluminum use in trains is more significant, particularly for freight cars but increasingly for passenger cars as well (Skillingberg and Green, 2007). Wrought alloys are used, particularly from the 5xxx and 6xxx series.

However, most of the aluminum used in transport applications goes into automobiles, sport utility vehicles (SUVs), light trucks, and motorcycles (Cooper, 2010). This use began after World War II and has increased steadily since. The average use of aluminum was 148 kg per vehicle in the United States in 2009 and is projected to reach 200 or even 250 kg per vehicle by 2025 (Gopalkrishnan, 2012). Automotive and other vehicles have a much longer life span than cans, so the effect of this increased usage on the scrap supply will not be noticed immediately. Nevertheless, it is projected that over 30% of all aluminum scrap will come from transport uses by 2040.

As the amount of aluminum has increased, the mix of alloys has changed. Nearly all of the early aluminum usage in automobiles was in the form of cast alloys, and cast alloys still comprise the vast majority of automotive aluminum (Cui and Roven, 2010; EAA, 2012). 3xxx alloys are widely used in powertrain components, particularly alloys 319, 356, 380, 383, and 384 (Gesing, 2005). Because of this, 56% of all cast-alloy aluminum production worldwide is used in automotive applications (Gopalkrishnan, 2012) and 75% in transport applications overall (Golden, 2009). However, the advantages of aluminum use in cars and trucks, particularly for weight reduction, create incentive for the use of wrought aluminum as well as cast alloys. Numerous applications use wrought alloys (Zapp et al., 2002), including bumpers (6061, 7003, 7108); seat frames (6061, 6063); *hang-on* parts like fenders, hoods, and deck lids (2022, 5182, 6016, 6111); radiators and condensers (1100, 3003, 4147); and structural parts like control arms and subframes (5754, 6016, 6061, 6082). The development of 6000-series alloys for panels is a particular focus of the industry (Sakurai, 2008). Almost 20% of extruded and rolled aluminum in Europe is now used for transport applications (Golden, 2009).

Common contaminants in transport-sector aluminum scrap include the following:

- Ferrous scrap, both attached (rivets) and separate pieces
- Magnesium and zinc die castings in automotive scrap
- Plastic and fabric in shredded automotive scrap

The influence of automotive aluminum use on recycling patterns is significant, since most recycled aluminum is used in this sector. The current predominance of cast alloys makes this easier, since cast alloys have a higher tolerance for impurities and can absorb a wider variety of scrap. This is particularly true of alloys 319.0 and 380.0 (Tessieri and Ng, 1995), making their employment in automotive applications even more attractive. However, the increasing use of wrought alloys may change this picture (Rombach, 2002), since the lower alloying-element content of wrought alloys makes them much less tolerant of mixed scrap loads. Because of this, the need exists for a means of separating aluminum scrap by alloy or alloy group. This has resulted in a series of technical advances in scrap processing with significant potential. These will be discussed in Chapter 5.

RECOMMENDED READING

Altenpohl, D.G., *Aluminum: Technology, Applications, and Environment*, 6th edn., TMS-AIME, Warrendale, PA, 1998, Chapters 14 and 15.

Kaufman, J.G., *Introduction to Aluminum Alloys and Tempers*, ASM International, Materials Park, OH, 2000, Chapter 6. http://www.pa–international.com.au/images/stories/Applications–for–Aluminum–Alloys–and–Tempers.pdf.

Menzie, W.D., Barry, J.J., Bleiwas, D.I., Bray, E.L., Goonan, T.G., and Matos G., The global flow of aluminum from 2006 through 2025, USGS Open-File Report 2010–1256, 2010. http://pubs.usgs.gov/of/2010/1256/pdf/ofr2010-1256.pdf.

REFERENCES

Altenpohl, D.G., *Aluminum: Technology, Applications, and Environment*, 6th edn., TMS-AIME, Warrendale, PA, 1998, Chapters 14 and 15.

Aluminum Association, Aluminum products: Electrical, June 28, 2012. http://www.aluminum.org/Content/NavigationMenu/TheIndustry/Electrical/default.htm.

Bijlhouwer, F., Why composition limits of popular extrusion alloys form an increasing obstacle for aluminium recycling, *Mater. Technol.*, 24, 157, 2009.

CAMA (Canadian Appliance Manufacturers Association), Generation and diversion of white goods from residential sources in Canada, January 2005. http://www.nrcan.gc.ca/sites/www.nrcan.gc.ca.minerals-metals/files/pdf/mms-smm/busi-indu/rad-rad/pdf/can-whit-goo-rec-05-eng.pdf.

Cayless, R.B.C., Alloy and temper designation systems for aluminum and aluminum alloys, in *Metals Handbook, Vol. 2: Properties and Selection: Nonferrous Alloys and Special-Purpose Materials*, 10th edn., ASM International, Materials Park, OH, 1990, p. 15.

Cooper, D.R., Steel and aluminium intensive products: Their metallic components and design requirements, July 2010. http://www.lcmp.eng.cam.ac.uk/wp-content/uploads/W6-Steel-and-aluminium-products.pdf.

Cui, J. and Roven, H.J., Recycling of automotive aluminum, *Trans. Nonferrous Met. Soc. China*, 20, 2057, 2010.

Das, S.K., Emerging trends in aluminum recycling, in *Aluminum Recycling and Processing for Energy Conservation and Sustainability*, Green, J.A.S., Ed., ASM International, Materials Park, OH, 2007, Chapter 9. http://www.asminternational.org/content/ASM/StoreFiles/05217G_Chapter09.pdf.

Das, S.K., Green, J.A.S., and Kaufman, J.G., Aluminum recycling: Economic and environmental benefits, *Light Metal Age*, 68(1), 22, 2010.

EAA (European Aluminium Association), The aluminium automotive manual: Materials—Resources, January 16, 2012. http://www.alueurope.eu/wp-content/uploads/2012/01/AAM-Materials-1-Resources.pdf.

Evermore Recycling, Management process/procedures manual: UBC product quality specification, September 5, 2012. http://www.alcoa.com/evermore/en/Documents/EMR_Product_Quality_Specification.pdf.

Froelich, D., Haoues, N., Leroy, Y., and Renard, H., Development of a new methodology to integrate ELV treatment limits into requirements for metal automotive part design, *Miner. Eng.*, 20, 891, 2007.

Gaustad, G., Olivetti, E., and Kirchain, R., Design for recycling: Evaluation and efficient alloy modification, *J. Ind. Ecol.*, 14, 286, 2010.

Gesing, A. and Wolanski, R., Recycling light metals from end-of-life vehicles, *JOM*, 53(11), 21, 2001.

Gesing, A.J., Aluminium, recycling and transportation, presented at *Aluminium 2005*, Kliczow, Poland, October 13, 2005. http://www.gesingconsultants.com/publications/50926.pdf.

Golden, C., *Aluminium Recycling: An International Review*, Imperial College London, London, U.K., 2009. http://warrr.org/785/1/13.27_IC_C_Golden_Masters_Thesis_2009.pdf.

Gopalkrishnan, S., Aluminum castings—The implications of strong growth in the coming decade, presented at *20th International Recycled Aluminium Conference*, Salzburg, Austria, November 21, 2012.

IAI (International Aluminium Institute), Global aluminium recycling: A cornerstone of sustainable development, October 29, 2009. http://www.world-aluminium.org/media/filer/2012/06/12/fl0000181.pdf.

Inada, T., Trend of food packaging materials in Japan, October 26, 2010. http://www.jpi.or.jp/english/packaging.htm.

JAA (Japan Aluminium Association), Outline of the Japanese aluminium industry, December 20, 2010. http://www.docstoc.com/docs/111235451/Auto-Industry-Forcast-2009.

Jones, M.R., Aluminium in the construction industry, *Kawneer White Paper 1999*, 2002. http://www.kawneer.com/kawneer/united_kingdom/en/pdf/Aluminum_in_the_Construction_Industry.pdf.

Kaufman, J.G., *Introduction to Aluminum Alloys and Tempers*, ASM International, Materials Park, OH, 2000, Chapter 6. http://www.pa–international.com.au/images/stories/Applications–for–Aluminum–Alloys–and–Tempers.pdf.

Kevorkijan, V., Advances in recycling of wrought aluminum alloys for added value maximization, *Metall. Mater. Eng.*, 16, 103, 2010. http://www.metalurgija.org.rs/mjom/Vol16/No2/3_Kevorkijan_MJoM_1602.pdf.

Kooiman, J., Aluminum vs. copper conductors: A serious alternative? 2011. http://www.interstates.com/img/site_specific/uploads/Aluminum_vs_Copper.pdf.

McMillan, C.A., Moore, M.R., Keoleian, G.A., and Bulkley, J.W., Quantifying U.S. aluminum in-use stocks and their relationship with economic output, *Ecol. Econ.*, 69, 2606, 2010. http://www-personal.umich.edu/~skerlos/quantifying_mcmillan.pdf.

Menzie, W.D., Barry, J.J., Bleiwas, D.I., Bray, E.L., Goonan, T.G., and Matos G., The global flow of aluminum from 2006 through 2025, USGS Open-File Report 2010–1256, 2010. http://pubs.usgs.gov/of/2010/1256/pdf/ofr2010-1256.pdf.

Morrison, J., European aluminium recycling under threat?, Alum. Int. Today, 17(1), 17, Jan. 2005.

Nishiyama, S., Aluminum can recycling in the systemized closed-loop, *Corr. Eng.*, 51, 285, 2002.

Rombach, G., Future availability of aluminium scrap, in *Light Metals 2002*, Schneider, W., Ed., TMS-AIME, Warrendale, PA, 2002, p. 1011.

Sakurai, T., The latest trends in aluminum alloy sheets for automotive body panels, *Kobelco Technol. Rev.*, (28), 22, 2008. http://www.kobelco.co.jp/english/ktr/pdf/ktr_28/022–028.pdf.

Skillingberg, M. and Green, J., Aluminum applications in the rail industry, *Light Metal Age*, 65(10), 8, 2007. http://www.aluminum.org/Content/NavigationMenu/TheIndustry/SheetPlate/Aluminum_Applications_in_the_Rail_Industry.pdf.

Southwire, Aluminum building wire 40 years later, October 11, 2012. http://www.southwire.com/commercial/AluminumBuildingWireHistory.htm.

Tessieri, M.B. and Ng, G.K., Forecast of aluminum usage in the automotive market and subsequent impact on the recycling infrastructure, in *Third International Symposium on Recycling of Metals and Engineered Materials*, Queneau, P.B. and Peterson, R.D., Eds., TMS-AIME, Warrendale, PA, 1995, p. 713.

Woehlke, J., Aluminum UBC's & evermore recycling, presented at *2011 BIR Autumn Round-Table Sessions*, Munich, Germany, October 24, 2011. http://webmail.bir.org/birweb/assets/private/presentations/Munich2011/NonFerrousWoehlke.pdf.

Zapp, P., Rombach, G., and Kuckshinrichs, W., The future of automotive aluminium, in *Light Metals 2002*, Schneider, W., Ed., TMS-AIME, Warrendale, PA, 2002, p. 1003.

3 Scrap Collection

MATERIALS LIFE CYCLE (HENSTOCK, 1996)

Figure 3.1 illustrates the *life cycle* in which raw materials are processed to generate consumer products and then disposed of after the life of those products ends. In the case of aluminum, the raw materials are the bauxite ore from which the metal is generated, the petroleum coke used to produce anodes and cathodes, and the other chemicals used in the Bayer process and the electrolytic cells. Primary production is the process by which aluminum is extracted from the bauxite, electrowon from a molten salt electrolyte, refined and cast into ingot or billet, and shipped to customers. The engineering materials generated by this process include pure aluminum and the alloys described in Chapter 2. Manufacturing is the process by which the ingot or billet is transformed into consumer products; as mentioned in Chapter 2, most aluminum placed in service is classified either as cast or wrought product. The types of products made from aluminum, and their usage patterns, have significant impacts on recycling technology and will be discussed in greater detail.

The scrap streams in Figure 3.1 are numbered according to their classification. Stream (1) is better known as *in-house, home,* or *run-around* scrap. In-house scrap consists of pure aluminum metal or alloy with a known composition, without any coatings or attachments. Ingot and billet croppings are the most common sources of in-house scrap, along with edge trimmings from sheet and plate. Because it is unadulterated, in-house scrap can be directly recycled simply by putting it back into the melting furnace. As a result, it is usually not included in recycling statistics. It is generated in both primary and secondary production plants. In addition, directly recycled manufacturing scrap (stream 2a) can be considered in-house scrap as well. Examples include casting gates and runners, skeletons from stamping operations, and croppings from extrusions. In-house scrap can usually be recycled directly back into the same alloy composition in which it was produced in the first place. This is a good example of *closed-loop* recycling.

Streams (2), (2′), and (in some cases) (2a) are known as *new, prompt industrial,* or *internal arising* scrap. Every manufacturing operation generates at least some new scrap, and some generate considerable amounts. Primary production, secondary production, and some manufacturing facilities also generate dross, which can be considered a form of new scrap as well. Table 3.1 lists the average percent of the input aluminum turned into scrap by manufacturing operations in different industrial sectors (Bruggink and Martchek, 2004). The values range from as little as 10% in foil production to 60% in airplane production. If the metal or alloy has not been adulterated, it can sometimes be recycled directly within the generating facility (stream 2a). Otherwise, it must be processed to a purity and form suitable for recycling and returned to a metal production facility. *Open-loop* recycling of this sort is the role of the secondary materials industries in Figure 3.1. The upgraded

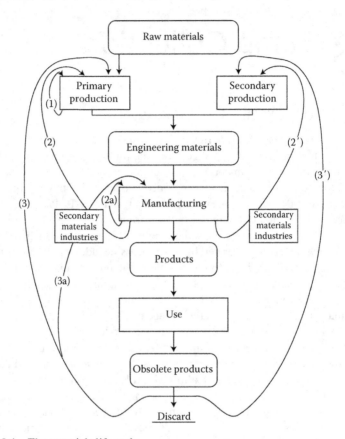

FIGURE 3.1   The materials life cycle.

scrap produced by the secondary materials industries is purchased by both primary (stream 2) and secondary (stream 2′) production facilities. In recent years, primary producers in industrialized countries have become more active in scrap purchasing.

Finally, streams (3), (3′), and (3a) are collectively known as *old, obsolete, post-consumer,* or *external arising* scrap. As the name suggests, old scrap consists of consumer products that have been discarded or taken out of service. The amount of old scrap available depends on several variables:

- *Urbanization*: In order for old scrap to be economically collected, it must be present in large enough amounts to justify the effort. As a result, the presence of abandoned cars, appliances, cans and bottles, etc., is much more apparent in rural areas.
- *Ease of recycling*: Large bulky items such as castings or extrusions are much more easily recycled than thin, heavily adulterated items such as aluminum foil and wiring. As a result, the former is more likely to be recycled. Table 3.2 shows the product life and recycling rate in the United States for aluminum products in various sectors (Bruggink and Martchek, 2004).

TABLE 3.1

Fabricator Recovery from Scrap by Market Sector

Market	Recovery (%)
Building and construction	80
Transportation—aerospace	40
Transportation—auto and light truck	75
Transportation—trucks, buses, and trailers	75
Transportation—rail	75
Transportation—other	75
Consumer durables	80
Electrical	90
Machinery and equipment	85
Containers and packaging—foil	90
Containers and packaging—other	75
Other	75

Source: Bruggink, P.R., in *4th International Symposium on Recycling of Metals and Engineered Material,* Stewart, D.L., Stephens, R., and Daley, J.C., Eds., TMS-AIME, Warrendale, PA, 2000. With permission.

The higher recycling rates are associated with transportation products, for which a well-developed recycling infrastructure exists.

- *Concentration*: Aluminum cans require more processing to recycle, but the percentage of aluminum in them is above 90%. The extrusions and aluminum conduit in building debris are easier to remelt, but the concentration of aluminum in the debris is about 0.01%. As a result, the recycling rate for aluminum cans is much higher than for aluminum from construction debris.
- *Legislation*: European Union regulations require the recycling of construction debris. As a result, the recycling rate for aluminum from buildings in Europe is 80% or higher (EAA, 2004), rather than the 15% US rate cited in Table 3.2 (Bruggink and Martchek, 2004). Providing a subsidy or other economic incentive also increases recycling rates, as demonstrated in Taiwan and the Netherlands (Lee, 1997; van Schaik et al., 2001).
- *Industrial involvement*: When industry deliberately acquires obsolete items of its own manufacture, the recycle rate goes up. Power and telecommunications firms have long collected scrap cable as they laid replacements. Can producers have constructed several facilities specifically for recycling UBCs (Woehlke, 2011) and now deliberately seek supplies of discarded cans (Miller, 2012). Other types of industries (car dealers, computer makers) are now required to take back their obsolete products (Muchová and Eder, 2010; Seebacher et al., 2006). (This is another reason why transportation sector recycle rates are high.) Lack of industry involvement in recovering obsolete items (e.g., *white goods,* the consumer durables in Table 3.2) results in a lower recycling rate.

TABLE 3.2

Product Life and Recycling Effectiveness for Market Types of Aluminum Scrap

Market	Average Product Life (Year)	Estimated Recycle Rate (%)	Estimated Metal Recovery (%)
Building and construction	40	15	85
Transportation— aerospace	30	30	90
Transportation—auto and light truck	13	80	90
Transportation—trucks, buses, and trailers	20	70	90
Transportation—rail	30	70	90
Transportation—other	20	70	90
Consumer durables	15	20	90
Electrical	35	10	90
Machinery and equipment	25	15	90
Containers and packaging—foil	1	2	80
Containers and packaging—other	1	25–60	90
Other	15	20	90

Source: Bruggink, P.R., in *4th International Symposium on Recycling of Metals and Engineered Material,* Stewart, D.L., Stephens, R., and Daley, J.C., Eds., TMS-AIME, Warrendale, PA, 2000. With permission.

Obsolete aluminum products that are not recycled are landfilled, lost, or *dissipated*. A dissipative use of aluminum is one that destroys its value as a potentially recyclable material. The best example is the use of aluminum as a deoxidant in molten steel. Low-grade aluminum scrap added to the steel reacts with dissolved oxygen, resulting in deoxidized steel and aluminum oxide dissolved in the slag. The value of the steel is increased, but the value of the aluminum is destroyed. Other dissipative uses include the use of aluminum as an alloying agent in bronzes and zinc die-casting alloys and the use of aluminum as a reducing agent in the metallothermic production of refractory metals and ferroalloys. The fraction of aluminum used in dissipative purposes is a very small one, however.

SCRAP COLLECTION PRACTICE

NEW SCRAP

The traditional approach to collecting new scrap involves three parties—the *dealer,* the *broker,* and the *processor* (Aluminum Association, 1998).

The scrap dealer is the primary acquirer of metal scrap from industrial sources (Minter, 2010). This involves entering into a contract to purchase the scrap, physically taking it to a central collection facility (better known as a *salvage yard, scrap yard,* or *junkyard*), and then reselling it to a broker or processor. The contract can be arranged either by direct negotiation with the generator or it can be the result of competitive bidding. Several means of transport can be used, but the relatively small amounts of scrap generated by most sources means that trucking is the most common method.

The broker purchases scrap from several traders (Gianotti, 2012; Minter, 2007, 2008b, 2010) and sells it in large lots to processors (Logue, 2009). By purchasing from several traders, the broker can act as a supermarket, offering customers different grades of scrap at different prices. This allows the broker to sell to each processor the type of scrap that best suits the particular processor's operation.

The processor converts scrap to a form that can be sold to a smelting or remelting facility (Minter, 2008a, 2011). Tasks that a processor may perform include

- Sorting the scrap to separate one metal from another or to separate different grades of crap from each other
- Shearing and/or shredding the scrap to *liberate* the aluminum from attachments and reducing it to a size convenient for transportation and further processing
- Cleaning the scrap to remove dirt and other extraneous material
- Baling or briquetting the scrap to make subsequent transport easier
- Simple melting (e.g., sweat melting) to separate aluminum in the scrap from attached iron and converting the scrap to ingot for shipping

In practice, these three roles are often not distinct from one another. Dealers frequently perform some processing and often sell their scrap directly to processors or smelters, bypassing the broker. Processors often negotiate directly with large scrap generators for scrap, eliminating the need for a dealer or a broker. This process of *vertical integration* has accelerated in the past 20 years.

A second change in new scrap collection and disposition involves the relationship between generator and processor. Increasingly, large generators of scrap deal directly with processors (Douglas, 2012; Gianotti, 2012), signing long-term contracts to supply all their scrap to the processors. The processors then sort and remelt the scrap, returning the secondary metal to the original generator. This eliminates the trader and broker and in some cases the scrap yard as well. In several cases, the scrap processors have set up operations adjacent to or on the generator's property, resulting in a dedicated recycling facility. This reduces costs and also improves the potential for closed-loop recycling, since the alloy content of the scrap sent to the dedicated recycling facility is known and can be easily segregated.

A third major change in scrap dealing and processing concerns the structure of the industry (Seabrook, 2008). Scrap dealers often began as small family-owned businesses, acquiring most of their scrap from the immediate region. However, better communications (in particular the Internet) have made it possible in recent years for dealers and processors to seek out scrap supplies from a much broader region (Gianotti, 2012), resulting in increased competition and decreased profit margins for smaller operations. In addition, small operations face increasing regulatory and environmental costs. Scrap yards located in urban areas are often targeted for closure under zoning laws (Gordon, 2012; Meyer, 2005), and the noise and dust that scrap processing generates are increasingly unacceptable. As a result, consolidation in the scrap trading and processing industry has been an ongoing trend. The family-owned scrap businesses have been sold to or merged with large public corporations that operate on a national or even global scale. Such large corporations can effectively fill the roles of trader, broker, and processor by themselves.

OLD SCRAP

As mentioned earlier, the sources of old scrap can be divided into the main use sectors of aluminum—transportation, buildings, packaging, wire and cable, and other.

Transportation: The collection of automotive scrap varies in different parts of the world. In general, end-of-life vehicles are obtained by one of four agents (Kumar and Sutherland, 2008; Woodyard, 1999):

- Automobile dealers, who obtain the old vehicles when customers trade them in for newer ones
- Repair shops, who receive the remains of vehicles that have been irreparably damaged by accident
- Highway authorities, who recover cars that have been abandoned by the roadside
- Salvage yard operators, who occasionally purchase vehicles directly from their owners (Paul, 2007)

The relative importance of these four in collecting depends on location, the perceived value of the car as scrap, and the regulatory environment. In rural environments, salvage yards are not as convenient to car owners, and the distance of the salvage yards from markets for the car or its parts make it less valuable. As a result, abandonment of obsolete vehicles is a more attractive option to owners (Gesing, 2005), and highway departments play a larger role in collection. In urban environments, salvage yards are more accessible, and the parts of the car are more valuable. This results in fewer abandonments and more recovery of obsolete vehicles by dealers and salvage yards.

The composition of automobiles has also impacted the pattern of obsolete automobile collection. As the amount of aluminum and magnesium in cars has increased (see Chapter 2), their value to the recycling industry has increased as well. (The inclusion of catalytic converters with small but valuable amounts of platinum-group metals has also encouraged collection.) This makes it more likely that obsolete automobiles will be returned for recycling through proper channels, rather than simply being abandoned. Approximately 95% of all passenger vehicles sold in the United States are now recovered for recycling (Kumar and Sutherland, 2008). The percentages recovered in Europe and Japan are similar.

Once collected, most obsolete vehicles are ultimately sold or given to salvage yards. About 6,000 of these exist in the United States and 16,000 in Europe, usually run as small family operations (EAA, 2002; Kumar and Sutherland, 2008; Nakajima and Vanderburg, 2005). The salvage yards perform the first steps in the recycling process. These include removing hazardous or environmentally difficult materials from the car, including fluids, glass, air bags, and mercury switches (Gesing, 2005; Paul, 2007). This is followed by dismantling those parts of the car that need to be recycled separately; these typically include the battery, copper radiators, catalytic converter, tires and wheels, and the gasoline tanks. About 15% of the aluminum in the vehicle is recovered during the dismantling phase. Salvage yards also remove parts that can be reused as replacement parts for other vehicles being repaired. These can include engines, transmissions, doors, bumpers, headlamps, fenders, and several other parts. This type of dismantling is done automatically in urban salvage yards and on a more haphazard basis in rural yards. Some salvage yards have begun to move *upstream* and have installed their own shredders (see Chapter 5), creating a more vertically integrated operation (Taylor, 2012).

As Table 3.2 shows (Bruggink and Martchek, 2004), recycling rates for buses and trucks are similar to those for personal vehicles. The lower rate for aerospace is misleading, since obsolete aircraft are more likely to be used as sources of parts than other vehicles, and as a result are not listed as recycled.

PACKAGING

By far the most common source of old aluminum scrap is packaging, in particular the UBC. As Chapter 2 points out, the beverage can has risen to prominence only since the mid-1960s. As a result, for some time, concerted efforts to recover the cans were not considered important. Descriptions of aluminum recycling technology from the 1960s and early 1970s treat UBCs as a minor source of metal and an undesirable one at that (Fine et al., 1973). However, this changed in the early 1970s, due to several reasons:

- Governments, noticing the increasing prevalence of aluminum cans in roadside litter, began enacting legislation banning or restricting the use of aluminum beverage containers. A common legislative initiative was to require payment of a deposit on each beverage container purchased and refundable when the empty container was returned (Gitlitz, 2003; Woehlke, 2011). As a response, the aluminum industry began developing infrastructure and technology specifically for collecting and recycling UBCs (Broughton, 1994).
- The energy crisis of the early 1970s had a significant impact on the cost of producing primary aluminum. The greatly reduced energy requirement for recycling metal made this alternative look increasingly attractive.
- Since an increasing percentage of the aluminum industry's output consisted of beverage cans, an increasing percentage of the available scrap consisted of this material too. If the industry was to obtain sufficient supplies of scrap, recycling UBCs was necessary. As a result, the US aluminum industry has

in recent years come to favor deposit legislation (Gardner, 2008). About 40% of the UBC supply in the United States currently comes from cans returned for the deposit (Leahy, 2010).

As a result, several pathways have been developed for the collection of UBCs. The higher rate of aluminum can usage in North America has meant greater development of collection infrastructure there. However, many of the same approaches are used elsewhere in the world as well. Collection methods include:

- *Part-time* and *subsistence collectors*, who gather aluminum cans from trash bins and other disposal areas (Miller, 2012).
- *Buy-back centers* sponsored by aluminum producers or recycling facilities (Green, 2006).
- *Municipal recycling facilities (MRFs)*, which sort through garbage to recover UBCs and other recyclables (Schaffer, 2009a).
- *Curbside separation and collection*, which feeds a separated stream of recyclables into the MRF (Jenkins et al., 2003). Curbside serves nearly half of the US population.

Individual collectors have a variety of economic and social backgrounds and various motivations (Miller, 2012). Most collect cans for the income (Khullar, 2009; Minter, 2010), some for the exercise, and some out of a sense of obligation toward the environment. Part-time collecting is seasonal (Harler, 1999), with far more cans being collected during the warmer months.

Buy-back centers differ considerably in their degree of sophistication and size. The simplest are first-tier facilities, which purchase cans and perform some rudimentary sorting (Aluminum Association, 1998). Figure 3.2 illustrates a typical reverse vending machine, which returns a small amount of cash for cans fed into it. Machines like this are typically employed in states or countries where mandatory deposits are required. Because of this, they can often accept several types of UBC in addition to aluminum and can perform sorting and flattening as well as collecting operations. Where mandatory deposits are not required, a first-tier facility may consist of nothing more than a drop-off trailer. Cans are left at the trailer and periodically retrieved by the site owner or manager.

When the number of cans being returned is sufficient to justify it, a second-tier facility will be constructed to process cans collected from first-tier facilities (Broughton, 1994; Leahy, 2010). Second-tier facilities have a magnetic sorter to remove steel cans and other ferrous scrap from the aluminum cans. They also have a baler or flattener to increase their bulk density and make the cans easier to ship. If the can collection operation is large enough, third-tier centers may be built (Aluminum Association, 1998). These are primarily processing rather than collection centers and perform shredding and cleaning of the collected cans in addition to magnetic separation.

Municipalities often encourage or require households and businesses to presort their trash before disposal. A common approach is the provision of a *blue bag*, into which recyclables are placed. Presorting can occur at three levels. The first is multiple-stream recycling, in which recyclables are separated into several bins.

FIGURE 3.2 A reverse vending machine. (From Can and Bottle Systems, Inc., http://www. canandbottle.com/pages/onestop.html. With permission.)

The list of accepted recyclables varies, but always includes metal scrap. Cardboard, glass, paper, and plastics may also have their own separate blue bag. A more common scheme is *dual-stream* recycling, which provides one blue bag for containers (including aluminum cans) and a second for other recyclables. The third type of presort is *single-stream* recycling, in which all recyclables are placed into a single blue bag. The contents are then either sorted at curbside or at an MRF. Figure 3.3

FIGURE 3.3 Sorting truck for unseparated recyclables. (From Kann Manufacturing Corporation, http://www.kannmfg.com/products/recycling/vh/. With permission.)

illustrates a truck specifically designed for curbside sorting. The type of collection system used is determined by the need to increase participation; the trend is toward adoption of the single-stream system (Schaffer, 2009b). However, UBCs collected by single-stream recycling have higher impurity levels, requiring more downstream processing (Leahy, 2010). About 20% of the UBC supply in the United States is obtained through curbside programs.

Alternatively, the municipality may choose to collect its garbage in one container and then separate out the recyclables at an MRF. The reasons for recovering recyclables are twofold (Lave et al., 1999; Malloy, 1998):

- Recovering recyclables reduces the amount of garbage to be landfilled, reducing land usage and cost of acquiring and operating landfills. In regions where available land is scarce (e.g., Japan), this is particularly important (Itou, 1995).
- The recovered recyclables have value. Many MRFs were constructed on the assumption that the proceeds from selling these recyclables (along with the savings from reduced landfill usage) would more than pay the cost of operating the MRF, becoming a net money earner for the operator. The aluminum cans in the garbage are particularly important in this regard, since they represent more than 20% of the total value of the recyclables, despite being less than 2% of the garbage (Aluminum Association, 1998; Das and Hughes, 2006).

In practice, MRFs have not always been as successful as originally envisioned (Reid, 2002). Low prices for recyclables, and lower-than-anticipated landfilling costs, have made many MRFs in North America unprofitable, and some have been closed or curtailed. Elsewhere, conditions have been more favorable.

As a result of these collection efforts, UBC recycling rates have risen steadily wherever concerted collection efforts have been made. Figure 3.4 shows the current

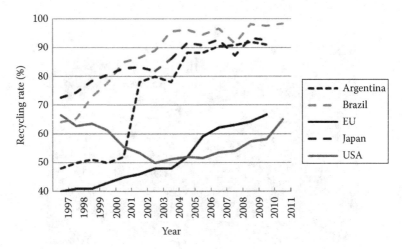

FIGURE 3.4 UBC recycling rates around the world, 1997–2011.

percentages in various areas of the world. Collection rates are currently highest in Brazil and Japan, at well over 80%. The European rate masks considerable differences between countries. In northern Europe, mandatory deposit legislation has pushed recycling rates to levels above 70% (Gitlitz, 2003; Green, 2006; Woehlke, 2011).

A note of concern can be found in the UBC recycling statistics for the United States shown in Figure 3.4. From a high of 67% in the early 1990s, the percentage of cans produced by the US aluminum industry that were processed for recycling fell to just over 50% in 2002 (Woehlke, 2011). Reasons for this include (Gitlitz, 2003):

- The low price of aluminum. This is reflected in a lower price for aluminum scrap, which means reduced incentive for collectors to recover cans.
- Reduced levels of public interest—recycling has in some ways become yesterday's news.
- The reduced cost of alternatives—with landfilling costs less than anticipated, and MRF operating costs greater than forecast, some municipalities no longer operate MRFs. The result is reduced UBC collection.
- The reduced mass of the can, which weighs one-third less than in 1970. This means that 33 cans are now required for a pound of metal instead of the 21 needed 30 years ago. This further reduces the incentive for collecting the cans.

Some of these factors will have a long-term impact on can collection rates. As a result, increasing these rates may require legislative initiatives, such as the mandatory deposit legislation described earlier.

Aluminum foil is also used in food and beverage packaging, especially aseptic *juice boxes*. These boxes are popular packages for single-serving products. However, separating the thin layer of aluminum foil from the layers of paper and plastic comprising the package is a complex and expensive operation (Steeman, 2013), difficult to justify given the negligible amount of metal recovered. As a result, the recycle rate for this material is only 2% in the United States, somewhat higher elsewhere.

WIRE AND CABLE

The fraction of recycled aluminum obtained from scrap wire and cable is likely no more than 5% (Bruggink and Martchek, 2004). Reasons for this include the relatively small amount of aluminum used in this application and the low recovery rate (see Table 3.2). Most aluminum-containing wires are collected from industrial sources: wire manufacturers, telephone companies, power and utility companies, and scrap dealers sorting building debris (Muchová and Eder, 2010).

The growing amount of electronic scrap generated by economically advanced societies points to another potential source of recycled aluminum. Printed circuit boards (PCB) consist of up to 7% metallic aluminum (Cui and Forssberg, 2006), and the overall aluminum content of waste electronic equipment is estimated at 4.7% (Muchová and Eder, 2010). Processes have been developed for recovering an aluminum *concentrate* as one of the products of circuit-board recycling. However, the relative value of the aluminum in a PCB is minimal compared with that of several other materials in the PCB, and as a result, actual aluminum recovery is small. The aluminum recovered from electronic materials is a very small fraction of the total old scrap collection (Bruggink and Martchek, 2004).

BUILDING AND CONSTRUCTION

Practices for recovery of aluminum from demolished or renovated buildings vary considerably between continents. In North America, recovery rates are low, perhaps about 15% (Bruggink and Martchek, 2004; Nisbet et al., 2003). The reason may be the nature of aluminum use in buildings, where it is used in assemblies that are difficult to dismantle. Without a regulatory imperative, construction and demolition (C&D) debris is more often landfilled than processed in North America.

The situation is different in Europe, where aluminum recovery rates from C&D debris are well over 90% (Das et al., 2010; EAA, 2004). Nonresidential buildings average 2% aluminum, residential buildings less than 1%; however, this could increase over time as the use of aluminum in residential constructions grows. Contrary to North American experience, much of the aluminum used in construction in Europe seems to be *hang-on* parts such as siding and window frames, easily removed before demolition. A regulatory regime that encourages/requires materials recovery from demolition and renovation projects also contributes to the high recovery rate.

RECOMMENDED READING

Aluminum Association, *Aluminum Recycling Casebook,* Aluminum Association, Washington, DC, 1998.

Bruggink, P.R. and Martchek, K.J., Worldwide recycled aluminum supply and environmental impact model, in *Light Metals 2004,* Tabereaux, A.T., Ed., TMS-AIME, Warrendale, PA, 2004, p. 907.

EAA (European Aluminium Association), *Collection of Aluminium from Buildings in Europe,* 2004. http://www.greenbuilding.world-aluminium.org/uploads/media/1256565286Delft_Brochure_and_TU_Delft_Report-1.pdf.

Kumar, V. and Sutherland, J.W., Sustainability of the automotive recycling infrastructure: Review of current research and identification of future challenges, *Int. J. Sust. Manuf.* 1, 145, 2008.

Seabrook, J., American scrap: An old–school industry globalizes, *The New Yorker,* 84(2), 46, January 14, 2008.

Woehlke, J., Aluminum UBC's & evermore recycling, presented at *2011 BIR Autumn Round–Table Sessions,* October 24, 2011. http://webmail.bir.org/birweb/assets/private/presentations/Munich2011/NonFerrousWoehlke.pdf.

REFERENCES

Aluminum Association, *Aluminum Recycling Casebook,* Washington, DC, 1998.

Broughton, A.C., A tradition of recycling, *Recycling Today,* 32(10), 36, 1994.

Bruggink, P.R., Aluminum scrap supply and environmental impact model, in *4th International Symposium on Recycling of Metals and Engineered Material,* Stewart, D.L., Stephens, R., and Daley, J.C., Eds., TMS-AIME, Warrendale, PA, 2000, p. 809.

Bruggink, P.R. and Martchek, K.J., Worldwide recycled aluminum supply and environmental impact model, in *Light Metals 2004,* Tabereaux, A.T., Ed., TMS-AIME, Warrendale, PA, 2004, p. 907.

Can and Bottle Systems, Inc., CBSI Onestop RUM, http://www.canandbottle.com/pages/onestop.html.

Cui, J. and Forssberg, E., Recycling of consumer electronic scrap, in *Proceedings of International Seminar on Mineral Processing Technology,* Bhaskar Raj, G., Ed., Allied Publishers Pvt. Ltd., Chennai, India, 2006, p. 14.

Das, S.K., Green, J.A.S., Kaufman, J.G., Emadi, D., and Mahfoud, M., Aluminum recycling—An integrated, industry wide approach, *JOM,* 62(2), 23, February 2010.

Das, S.K. and Hughes, M., Improving can recycling rates: A six sigma study in Kentucky, *JOM,* 58(8), 27, August 2006.

Douglas, A., Global scrap flows and consumption patterns, presented at *Metal Bulletin 20th International Recycled Aluminium Conference,* Salzburg, Austria, November 20, 2012.

EAA (European Aluminium Association), *The Aluminium Automotive Manual,* 2002. http://www.eaa.net/aam.

EAA (European Aluminium Association), *Collection of Aluminium from Buildings in Europe,* 2004. http://www.greenbuilding.world-aluminium.org/uploads/media/1256565286Delft_Brochure_and_TU_Delft_Report-1.pdf.

Fine, P., Rasher, H.W., and Wakesburg, S., *Operations in the Nonferrous Scrap Metal Industry Today,* NASMI, New York, 1973, p. 25.

Gardner, S., The Aluminum Association announces recycling target, November 18, 2008. http://www.aluminum.org/AM/Template.cfm?Section=Home&CONTENTID=27264&TEMPLATE=/CM/ContentDisplay.cfm.

Gesing, A.J., Aluminium, recycling and transportation, presented at *Aluminium 2005,* Kliczkow Castle, Poland, October 13, 2005. http://www.gesingconsultants.com/publications/50926.pdf.

Gianotti, D., Scrap flows: Traders, producers & closed loops, presented at *Metal Bulletin 20th International Recycled Aluminium Conference,* Salzburg, Austria, November 20, 2012.

Gitlitz, J., *The Role of the Consumer in Reducing Primary Aluminum Demand,* Container Recycling Institute, Sao Luis, Brazil, 2003. http://www.container-recycling.org/assets/pdfs/aluminum/Aluminum-RoleofConsumer.pdf.

Gordon, L., Full of scrap: Eminent domain efforts sell scrap yards down the river, *Am. Metal Market,* 121(12), 14, December 2012.

Green, J., Recyclable aluminum rolled products, *Light Metal Age,* 64(8), 33, August 2006.

Harler, C., Stealing aluminum, *Recycling Today,* 37(3), 40, March 1999.

Henstock, M.E., *The Recycling of Non–Ferrous Metals,* ICME, Ottawa, Canada, 1996, pp. 51–56.

Itou, T., Recycling of used aluminum beverage cans in Japan, in *3rd International Symposium on Recycling of Metals and Engineered Materials,* Queneau, P.B. and Peterson, R.D., Eds., TMS-AIME, Warrendale, PA, 1995, p. 703.

Jenkins, R.R., Martinez, S.A., Palmer, K., and Podolsky, M.J., The determinants of household recycling: A material–specific analysis of recycling program features and unit pricing, *J. Environ. Econ. Manag.*, 45, 294, 2003.

Kann Manufacturing Corporation, Kann Versa-Haul, 13 August 2013, http://www.kannmfg.com/products/recyling/vh/

Khullar, M., Surviving on scrap, *Scrap*, 66(5), 58, September/October 2009.

Kumar, V. and Sutherland, J.W., Sustainability of the automotive recycling infrastructure: Review of current research and identification of future challenges, *Int. J. Sust. Manuf.*, 1, 145, 2008.

Lave, L.B., Hendrickson, C.T., Conway–Schempf, N.M., and McMichael, F.C., Municipal solid waste recycling issues, *J. Environ. Eng.*, 125, 944, 1999.

Leahy, M., Evermore recycling, presented at *Arizona Recycling Coalition & American Public Works Association Annual Conference*, Phoenix, AZ, August 3, 2010. http://www.arizonarecyclingcoalition.com/2010_conference/presentations/Evermore Recycling.pdf.

Lee, C.-H., Management of scrap car recycling, *Resour. Conserv. Recycl.*, 20, 207, 1997.

Logue, A.C., The last of the independents? *Scrap*, 66(6), 59, November/December 2009.

Malloy, M.G., Making the (higher) grade, *Waste Age*, 29(9), 57, September 1998.

Meyer, H., Masters of their (eminent domain), *Scrap*, 62(6), 57, November/December 2005.

Miller, J.W., The aluminum can wars begin, *Wall Street J.*, 25 September 2012; http://online.wsj.com/article/SB10000872396390443589304577633410750041328.html.

Minter, A., Sultans of scrap, *Scrap*, 64(3), 82, May/June 2007.

Minter, A., Scrap made in China, *Scrap*, 65(2), 154, March/April 2008a.

Minter, A., Thailand's taste for scrap, *Scrap*, 65(5), 80, September/October 2008b.

Minter, A., Brazil rising, *Scrap*, 67(3), 58, May/June 2010.

Minter, A., The industrial revolution, *Scrap*, 68(2), 142, March/April 2011.

Muchová, L. and Eder, P., End–of–waste criteria for aluminium and aluminium alloy scrap: Technical proposals, JRC Scientific and Technical Reports, EUR 24396, 2010. http://ftp.jrc.es/EURdoc/JRC58527.pdf.

Nakajima, N. and Vanderburg, W.H., A failing grade for the German end-of-life vehicles take-back system, *Bull. Sci. Technol. Soc.*, 25(2), 170, April 2005.

Nisbet, M., Venta, G., and Foo, S., Demolition and deconstruction: Review of the current status of reuse and recycling of building materials, 25 March 2003. ftp://ftp.tech-env.com/pub/RETROFIT/awmapaper_wm1b.pdf.

Paul, R.T., The success of vehicle recycling in North America, in *Light Metals 2007*, Sørlie, M., Ed., TMS-AIME, Warrendale, PA, 2007, p. 1115.

Reid, R.L., What's ailing the aluminum can? *Scrap*, 59(5), 30, November/December 2002.

Schaffer, P., Why UBC consumers are finally warming up to MRFs, *Am. Metal. Market*, 118(2), 26, March 2009a.

Schaffer, P., Single–stream recycling, *Am. Metal Market*, 118(2), 24, March 2009b.

van Schaik, A., Dalmijn, W.L., and Reuter, M.A., Impact of economy on the secondary materials cycle, in *Waste Proceedings of Recycling Minerals Metallurgy Industry IV*, Rao, S.R. et al., Eds., CIM, Montreal, Canada, 2001, p. 407.

Seabrook, J., American scrap: An old–school industry globalizes, *The New Yorker*, 84(2), 46, January 14, 2008.

Seebacher, H., Sunk, W., Antrekowitsch, H., and Klade, M., Recycling von aluminium in der automobilindustrie, *Aluminium*, 82(1/2), 24, January/February 2006.

Steeman, A., Recycling packaging material with an aluminium component, January 7, 2013. http://bestinpackaging.com/2012/02/02/recycling-packaging-material-with-an-aluminium-component/

Taylor, B. In-house service, *Recycling Today*, 50(9), 44 September 2012. http://www.recyclingtoday.com/rt0912-u-pull-it-profile.aspx.

Woehlke, J., Aluminum UBC's & Evermore Recycling, presented at *2011 BIR Autumn Round–Table Sessions*, October 24, 2011. http://webmail.bir.org/birweb/assets/private/presentations/Munich2011/NonFerrousWoehlke.pdf.

Woodyard, M., French makers take different recycling routes, *Automot. News,* 74, 30J, November 22, 1999.

4 Aluminum Recycling Economics

Although public support for recycling as a means of achieving social goals is widespread, in the end recycling is not a religious activity. It is not carried out by people who are trying to save the planet. Recycling is carried out by people and organizations who are trying to earn a profit. They face competition from other recyclers and from producers of primary material and do not control the price of their product. Their success depends on their ability to purchase raw materials at the lowest possible price and process them at the lowest possible cost. As a result, the economics of recycling is an important part of the secondary aluminum story and one that is critical to understanding the choices of processing strategy described in the next few chapters. The discussion will include both the cost of what recyclers use and the value of what they produce.

SCRAP GRADES

Figure 4.1 lists the grades and specifications of aluminum-containing scrap compiled by the US Institute for Scrap Recycling Industries (ISRI, 2012). Different scrap grades have different values; a comparison of some of these grades will explain why.

The first two grades in Figure 4.1, Tablet and Tabloid, are largely identical. However, Tabloid is new scrap, free of ink and paint. Tablet is old scrap and thus has ink and possibly paint on it. As a result, Tabloid will have greater value than Tablet. Several other types of scrap have separate grades for new and old material; the new material is always worth more.

Taboo is a common grade of aluminum scrap (Figure 4.2). Its specification demands little or no alloyed copper. This increases the possibility of using it to produce wrought alloys when it is remelted. The tolerance problem was discussed in Chapter 2; the low-copper specification is a result of this. Taboo also has limitations on smaller-sized materials and the amount of oil and grease it can contain. Smaller scrap pieces are more likely to oxidize when put in a melting furnace, lowering overall yields and generating more dross. Excessive grease will generate smoke and soot during the melting process, damaging the workplace environment. A plant that only purchases oil-free scrap can avoid having to purchase a decoating furnace (see Chapter 5). Limitations of these types are common through the descriptions in Figure 4.1.

There are six types of UBC scrap—Take, Talc, Talcred, Taldack, Taldon, and Taldork. Take is new scrap (Figure 4.3); the others are old. Talc and Talcred are loose scrap; the others have been densified by briquetting, bundling, or baling. Densified

Tablet—Clean Aluminum Lithographic Sheets
To consist of alloys 1100 and/or 3003, to be free of paper, plastic, excessively inked sheets and any other contaminants. Minimum of three inches (8 cm) in any direction.

Tabloid—New, Clean Aluminum Lithographic Sheets
To consist of 1000 and/or 3000 series alloys, to be free of paper, plastic, ink, and any other contaminants. Minimum size of three inches (8 cm) in any direction.

Taboo—Mixed Low Copper Aluminum Clippings and Solids
Shall consist of new, clean, uncoated and unpainted low copper aluminum scrap of two or more alloys with a minimum thickness of 0.015 in. (0.38 mm) and to be free of 2000 and 7000 series, foil, hair wire, wire screen, punchings less than ½ in. (1.25 cm) diameter, dirt, and other nonmetallic items. Grease and oil not to total more than 1%. Variations to this specification should be agreed on prior to shipment between the buyer and seller.

Taint/Tabor—Clean Mixed Old Alloy Sheet Aluminum
Shall consist of clean old alloy aluminum sheet of two or more alloys, free of foil, Venetian blinds, castings, hair wire, screen wire, food or beverage containers, radiator shells, airplane sheet, bottle caps, plastic, dirt, and other nonmetallic items. Oil and grease not to total more than 1%. Up to 10% Talc permitted.

Take—New Aluminum Can Stock
Shall consist of new low copper aluminum can stock and clippings, clean, lithographed or not lithographed, and coated with clear lacquer but free of lids with sealers, iron, dirt and other foreign contamination. Oil not to exceed 1%.

Talc—Post-Consumer Aluminum Can Scrap
Shall consist of old aluminum food and/or beverage cans. The material is to be free of other scrap metals, foil, tin cans, plastic bottles, paper, glass, and other nonmetallic items. Variations to this specification should be agreed upon prior to shipment between the buyer and seller.

Talcred—Shredded Aluminum UBC Scrap
Shall have a density of 12–17 pounds per cubic foot (193–273 kg/m³). Material should contain maximum 5% fines less than 4 mesh (US standard screen size) (6.35 mm). Must be magnetically separated material and free of steel, lead, bottle caps, plastic cans and other plastics, glass, wood, dirt, grease, trash, and other foreign substances. Any free lead is basis for rejection. Any and all aluminum items, other than used beverage cans, are not acceptable. Variations to this specification should be agreed upon prior to shipment between the buyer and seller.

Taldack—Densified Aluminum UBC Scrap
Shall have a biscuit density of 35–50 pounds per cubic foot (562–802 kg/m³). Each biscuit not to exceed 60 pounds (27.2 kg). Nominal biscuit size range from 10″ to 13″ × 10¼″ (25.4 × 33 × 26 cm) to 20″ × 6¼″ to 9″ (50.8 × 15.9 × 22.9 cm). Shall have banding slots in both directions to facilitate bundle banding. All biscuits comprising a bundle must be of uniform size. Size: Bundle range dimensions acceptable are 41″ to 44″ × 51″ (104–112 cm) to 54″ × 54″ (137 × 137 cm) to 56″ (142 cm) high. The only acceptable tying method shall be as follows: Using minimum 5/8″ (1.6 cm) wide by .020″ (0.05 cm) thick steel straps, the bundles are to be banded with one vertical band per row and a minimum of two girth (horizontal) bands per bundle. Use of skids and/or support sheets of any material is not acceptable. Must be magnetically separated material and free of steel, lead, bottle caps, plastic cans and other plastic, glass, wood, dirt, grease, trash, and other foreign substances. Any free lead is basis for rejection. Any and all aluminum items, other than UBCs, are not acceptable. Items not covered in the specifications, including moisture, and any variations in the specification should be agreed upon prior to shipment between the buyer and seller.

Taldon—Baled Aluminum UBC Scrap
Shall have a minimum density of 14 lbs per cubic foot (225 kg/m³), and a maximum density of 17 lbs per cubic foot (273 kg/m³) for unflattened UBC and 22 lbs per cubic foot (353 kg/m³) for flattened UBC.

FIGURE 4.1 Specifications for grades of aluminum scrap.

Size: minimum 30 cubic feet (0.85 m³), with bale range dimensions of 24″ to 40″ (61–132 cm) × 30″ to 52″ (76–132 cm) × 40″ to 84″ (102–213 cm). The only acceptable tying method shall be as follows: four to six 5/8″ (1.6 cm) × 0.20″ (5 mm) steel bands, or 6 to 10 #13 gauge steel wires (aluminum bands or wires are acceptable in equivalent strength and number). Use of skids and/or support sheets of any material is not acceptable. Must be magnetically separated material and free of steel, lead, bottle caps, plastic cans and other plastics, glass, wood, dirt, grease, trash and other foreign substances. Any free lead is basis for rejection. Any and all aluminum items, other than used beverage cans, are not acceptable. Variations to this specification should be agreed upon prior to shipment between the buyer and seller.

Taldork—Briquetted Aluminum UBC Scrap

Shall have a briquette density of 50 pounds per cubic foot (800 kg/m³) minimum. Nominal briquette size shall range from 12″ to 24″ (30.5–61 cm) × 12″ to 24″ (30.5–61 cm) in uniform profile with a variable length of 8″ (20.3 cm) minimum and 48″ (122 cm) maximum. Briquettes shall be bundled or stacked on skids and secured with a minimum of one vertical band per row and a minimum of one girth band per horizontal layer. Briquettes not to overhang pallet. Total package height shall be 48″ (122 cm) maximum. Banding shall be at least 5/8″ (1.6 cm) wide by .020″ (5 mm) thick steel strapping or equivalent strength. The weight of any bundle shall not exceed 4000 pounds (1.814 mt). Material must be magnetically separated and free of steel, plastic, glass, dirt, and all other foreign substances. Any and all aluminum items, other than UBC, are unacceptable. Any free lead is a basis for rejection. Items not covered in the specification, including moisture, and any variations to this specification should be agreed upon prior to shipment between the buyer and seller.

Tale—Painted Siding

Shall consist of clean, low copper aluminum siding scrap, painted one or two sides, free or iron, dirt, corrosion, fiber, foam, or fiberglass backing or other nonmetallic items.

Talk—Aluminum Copper Radiators

Shall consist of clean aluminum and copper radiators, and/or aluminum fins or copper tubing, free of brass tubing, iron and other foreign contamination.

Tall —E.C. Aluminum Nodules

Shall consist of clean E.C. aluminum, chopped or shredded, free of screening, hair-wire, iron, copper, insulation, and other nonmetallic items. Must be free of minus 20 mesh materials. Must contain 99.45% aluminum content.

Talon—New Pure Aluminum Wire and Cable

Shall consist of new, clean, unalloyed aluminum wire or cable free from hair wire, ACSR, wire screen, iron, insulation and other nonmetallic items.

Tann—New Mixed Aluminum Wire and Cable

Shall consist of new, clean unalloyed aluminum wire or cable, which may contain up to 10% 6000 series wire and cable free from hair wire, wire screen, iron, insulation, and other nonmetallic items.

Taste—Old Pure Aluminum Wire and Cable

Shall consist of old, unalloyed aluminum wire and cable containing not over 1% free oxide or dirt and free from hair wire, wire screen, iron, insulation, and other nonmetallic items.

Tassel—Old Mixed Aluminum Wire and Cable

Shall consist of old, unalloyed aluminum wire and cable, which may contain up to 10% 6000 series wire and cable with not over 1% free oxide or dirt and free from hair wire, wire screen, iron, insulation, and other nonmetallic items.

FIGURE 4.1 (continued) Specifications for grades of aluminum scrap.

(continued)

Tarry A—Clean Aluminum Pistons
Shall consist of clean aluminum pistons to be free from struts, bushings, shafts, iron rings and other nonmetallic items. Oil and grease not to exceed 2%.

Tarry B—Clean Aluminum Pistons with Struts
Shall consist of clean whole aluminum pistons with struts. Material is to be free from bushings, shafts, iron and other nonmetallic items. Oil and grease not to exceed 2%.

Tarry C—Irony Aluminum Pistons
Shall consist of aluminum pistons with non-aluminum attachments to be sold on a recovery basis or by special arrangement between buyer and seller.

Teens—Segregated Aluminum Borings and Turnings
Shall consist of aluminum borings and turnings of one specified alloy. Material should be free of oxidation, dirt, free iron, stainless steel, magnesium, oil, flammable liquids, moisture and other nonmetallic items. Fines should not exceed 3% through a 20 mesh (U.S. standard screen).

Telic—Mixed Aluminum Borings and Turnings
Shall consist of clean, uncorroded aluminum borings and turnings of two or more alloys and subject to deductions for fines in excess of 3% through a 20 mesh screen and dirt, free iron, oil, moisture and all other nonmetallic items. Material containing iron in excess of 10% and/or free magnesium or stainless steel or containing highly flammable cutting compounds will not constitute good delivery. To avoid dispute, material should be sold on basis of definite maximum zinc, tin and magnesium content.

Tense—Mixed Aluminum Castings
Shall consist of all clean aluminum castings which may contain auto and airplane castings but no ingots, and to be free of iron, brass, dirt, and other nonmetallic items. Oil and grease not to total more than 2%.

Tepid—Airplane Sheet Aluminum
Should be sold on recovery basis or by special arrangements with purchaser.

Terse—New Aluminum Foil
Shall consist of clean, new, pure, uncoated, 1000 and/or 3000 and/or 8000 series alloy aluminum foil, free from anodized foil, radar foil and chaff, paper, plastics, or other nonmetallic items. Hydraulically briquetted material and other alloys by agreement between buyer and seller.

Tesla—Post Consumer Aluminum Foil
Shall consist of baled old household aluminum foil and formed foil containers of uncoated 1000, 3000, and 8000 series aluminum alloy. Material may be anodized and contain a maximum of 5% organic residue. Material must be free from radar chaff foil, chemically etched foil, laminated foils, iron, paper, plastic and other nonmetallic contaminants.

Tetra—New Coated Aluminum Foil
Shall consist of new aluminum foil coated or laminated with ink, lacquers, paper, or plastic. Material shall be clean, dry free of loose plastic, PVC and other nonmetallic items. This foil is sold on a metal content basis or by sample as agreed between buyer and seller.

Thigh—Aluminum Grindings
Should be sold on recovery basis or by special arrangements with purchaser.

Thirl—Aluminum Drosses, Spatters, Spillings, Skimmings and Sweepings
Should be sold on recovery basis or by special arrangements with purchaser.

Throb—Sweated Aluminum
Shall consist of aluminum scrap that has been sweated or melted into a form or shape such as an ingot, sow or slab for convenience in shipping; to be free from corrosion, dross or any non-aluminum inclusions. Should be sold subject to sample or analysis.

FIGURE 4.1 (continued)   Specifications for grades of aluminum scrap.

Tooth—Segregated New Aluminum Alloy Clippings and Solids

Shall consist of new, clean, uncoated, and unpainted aluminum scrap of one specified aluminum alloy only with a minimum thickness of 0.015″ (0.38 mm) and to be free of hair wire, wire screen, dirt, and other nonmetallic items. Oil and grease not to total more than 1%. Also free from punchings less than ½″ (1.27 cm) in size.

Tough—Mixed New Aluminum Alloy Clippings and Solids

Shall consist of new, clean, uncoated and unpainted aluminum scrap of two or more alloys with a minimum thickness of 0.015″ (0.38 mm) and to be free of hair wire, wire screen, dirt, and other nonmetallic items. Oil and grease not to total more than 1%. Also free from punchings less than ½″ (1.27 cm) in size.

Tread—Segregated New Aluminum Castings, Forgings, and Extrusions

Shall consist of new, clean, uncoated aluminum castings, forgings, and extrusions of one specified alloy only and to be free from sawings, stainless steel, zinc, iron, dirt, oil, grease, and other nonmetallic items.

Troma—Aluminum Auto or Truck Wheels

Shall consist of clean, single-piece, unplated aluminum wheels of a single specified alloy, free of all inserts, steel, wheel weights, valve stems, tires, grease and oil, and other nonmetallic items. Variations to this specification should be agreed upon prior to shipment between the buyer and seller.

Trump—Aluminum Auto Castings

Shall consist of all clean automobile aluminum castings of sufficient size to be readily identified and to be free from iron, dirt, brass, bushings, and nonmetallic items. Oil and grease not to total more than 2%.

Twang—Insulated Aluminum Wire Scrap

Shall consist of aluminum wire scrap with various types of insulation. To be sold on a sample or recovery basis, subject to agreement between buyer and seller.

Tweak—Fragmentizer Aluminum Scrap (from Automobile Shredders)

Derived from either mechanical or hand separation, the material must be dry and not contain more than 4% maximum free zinc, 1% maximum free magnesium, and 1.5% maximum of analytical iron. Not to contain more than a total 5% maximum of nonmetallics, of which no more than 1% shall be rubber and plastics. To be free of excessively oxidized material, air bag canisters, or any sealed or pressurized items. Any variation to be sold by special arrangement between buyer and seller.

Twire—Burnt Fragmentizer Aluminum Scrap (from Automobile Shredders)

Incinerated or burned material must be dry and not contain more than X% (% to be agreed upon by buyer and seller) ash from incineration, 4% maximum free zinc, 1% maximum free magnesium, and 1.5% maximum of analytical iron. Not to contain more than a total 5% maximum of nonmetallics, of which no more than 1% shall be rubber and plastics. To be free of excessively oxidized material, air bag canisters, or any sealed or pressurized items. Any variation to be sold by special arrangement between buyer and seller.

Twist—Aluminum Airplane Castings

Shall consist of clean aluminum castings from airplanes and to be free from iron, dirt, brass, bushings, and nonmetallic items. Oil and grease not to total more than 2%.

Twitch—Floated Fragmentizer Aluminum Scrap (from Automobile Shredders)

Derived from wet or dry media separation device, the material must be dry and not contain more than 1% maximum free zinc, 1% maximum free magnesium, and 1% maximum of analytical iron. Not to contain more than a total 2% maximum of nonmetallics, of which no more than 1% shall be rubber and plastics. To be free of excessively oxidized material, air bag canisters, or any sealed or pressurized items. Any variations to be sold by special arrangement between buyer and seller.

FIGURE 4.1 (continued) Specifications for grades of aluminum scrap.

(continued)

Zorba—Recyclable Concentrates of Shredded Mixed Nonferrous Scrap Metal in Pieces—Derived from Fragmentizers for Further Separation of Contained Materials

Shall be made up of a combination of the nonferrous metals: aluminum, copper, lead, magnesium, stainless steel, nickel, tin, and zinc, in elemental or alloyed (solid) form. The percentage of each of these metals within the nonferrous concentrate shall be subject to agreement between buyer and seller, may vary from shredder to shredder and may, in some cases, be zero for a particular metal. Shall be obtained by air separation, flotation, screening, eddy current, other segregation technique(s) or a combination of the same. Shall have passed one or more magnets to reduce or eliminate free iron and/or iron attachments. Shall be free of radioactive material, dross, or ash. May be screened to permit description by specific size ranges. May contain high-density nonmetallics such as rock, glass, rubber, plastic, and wood. Items of exclusion, inclusion or limitation not set out in the above specifications, such as moisture and free iron and/or attachments or the presence or absence of other metals, are subject to agreement between buyer and seller. Material to be traded under this guideline shall be identified as ZORBA with a number to follow, indicating the estimated percentage nonferrous metal content of the material (e.g., ZORBA 63 – means the material contains approximately 63% nonferrous metal content).

FIGURE 4.1 (continued)   Specifications for grades of aluminum scrap. (From Institute of Scrap Recycling Industries [ISRI], *Scrap Specifications Circular 2012*, http://www.isri.org/iMIS15_Prod/ISRI/_Program_and_Services/Commodities/Scrap_Specifications/ISRI/_Program_and_Services/Scrap_Specifications_Circular.aspx?hkey=5c76eb15-ec00-480e-b57f-e56ce1ccfab5, 2012. With permission.)

material is easier to handle and is worth more. Talc does not specify magnetically separated materials; the other old UBC grades do. Scrap grades often specify required processing techniques, which produce improved purity and handling characteristics. Two of the newer scrap grades are EM-1 and EM-2, which specify the use of eddy current separation (see Chapter 5). Many scrap grades also specify a single product, such as UBCs, auto or truck wheels (Troma), pistons (Tarry), or automotive radiators (Tally).

FIGURE 4.2   Mixed low-copper clips (ISRI grade Taboo). (Photo courtesy of Ray Peterson, Aleris International, Beachwood, OH.)

FIGURE 4.3 New UBC scrap (ISRI grade Take). (Photo courtesy of Ray Peterson, Aleris International, Beachwood, OH.)

Two of the scrap grades in Figure 4.1 contain a mix of aluminum and other metals. These are Zeppelin and Zorba, which are products from the recycling of automobiles (see Chapter 6). Zorba is the nonferrous metallic fraction remaining when automobiles are shredded, passed under a magnetic separator to recover most of the iron and steel, and further processed to remove most of the nonmetallic material. Its composition varies, depending on the age and type of automobiles being shredded and the upgrading steps that it has gone through. Zeppelin is what remains of Zorba after it has been through a heavy media separator (see Chapter 5) to remove heavier metals such as copper, stainless steel, and zinc. It still contains some magnesium in addition to aluminum. When Zeppelin is further processed to remove magnesium, the result is Twitch. Much of the aluminum scrap traded worldwide is sold as these grades, particularly Twitch and Zorba.

Table 4.1 compares the ISRI scrap grade definitions with some of the definitions created by the German Verband Deutscher Metallhändler (VDM, 2011). The differences between the two sets of definitions are worth noting. The VDM grades include no specifications for UBC scrap, compared with six ISRI types. As mentioned previously, aluminum can usage is much greater in North America, resulting in much more scrap to process. On the other hand, VDM describes two grades of *rolled* scrap that have no direct ISRI equivalent. Rolled scrap includes plate, sheet, and foil (including UBCs). Because the North American scrap market is larger, ISRI can break this into several specific grades. Also included in Table 4.1 for comparison are commercially traded scrap grades defined by the industry publication Platts Metals Week (2012). These are the types of scrap common enough to have publicly listed prices.

TABLE 4.1

Comparison of ISRI, VDM, and Platts Aluminum Scrap Grades

ISRI Grade	VDM	Platts
Tablet		
Tabloid	Album/Ampel	
Taboo		US smelter-grade MLCCs[a]
Taint/Tabor		US old sheet
Take		
Talc		
Talcred		
Taldack		
Taldon		US UBCs
Taldork		
Tale		US painted siding
Talk		
Tall	Ahorn	
Tally		
Talon	Abweg	
Tann		
Tarry A		
Tarry B	Armin	
Tarry C	Artur	
Tassel	Adler	
Taste		
Tata	Alter	US 6063 press scrap
Toto		
Tutu		
Teens	Atoll	
Telic	Atlas/Autor	US turnings
Tense	Assel	
Tepid		
Terse	Arche	
Tesla		
Tetra		
Thigh		
Thirl	Azur	
Throb		
Tooth		
Tough		
Tread		
Troma		
Trump		
Twang		
Twist		
Twitch		US high-grade auto shreds
Tweak	Apsis	US low-grade auto shreds
Twire		

[a] Mixed low-copper clips.

SCRAP PRICES

Several factors play a role in determining the value of different scrap grades. Some of these are characteristics of the scrap itself, and some are external factors. Characteristics of the scrap that impact the price include the following:

- *Alloy purity*: Scrap that consists entirely of one alloy is more valuable than a mixture of alloys. It can be remelted directly to produce that alloy, rather than requiring blending with other alloys or dilution with primary metal.
- *Contaminants*: These include oil, grease, dirt, plastic, paper, and other metals. These must be removed prior to remelting, which adds to the processing cost. New scrap grades are worth more than old scrap in part because they have lower levels of contaminants.
- *Size*: Big pieces of scrap produce better yield when melted than finely divided scrap such as turnings and wire chops. As a result, they are worth more. Baled or briquetted pieces of small scrap are worth more than loose pieces, for the same reason.
- *Coatings and attachments*: These have to be removed, which increases the cost of processing. As a result, bare scrap is usually worth more.

External factors include the following:

- *The price of primary aluminum*: When primary prices go up, scrap prices go up as well, and when prices go down, scrap becomes cheaper as well. However, the difference between primary and scrap prices is not constant. It can be impacted by government policy (see Chapter 11), the cost of alloying elements, the amount of primary produced in a given geographical area, and short-term supply constraints (Xiarchos, 2005). When primary aluminum is plentiful, the gap between primary and scrap prices grows; when primary aluminum plants shut down (as many have in Europe), scrap becomes more valuable (Blomberg and Söderholm, 2009). Primary aluminum is always a *competitor* of recycled metal, and if the gap between the price of primary and that of scrap becomes too small, it becomes impossible to recycle metal at a profit. Margins in Europe shrank to nearly zero in 2009, as the global recession created a large surplus of primary metal, driving down prices (Stulgis, 2012).
- *The location of the scrap*: Scrap that is located a long way from a remelting or refining plant is more expensive to transport and thus less valuable. Table 4.2 shows the price of a single grade of scrap (segregated low-copper clips, usually alloy 5052; ISRI grade Taboo) at various locations in North America on a given day (AMM, 2012). The highest values are on the West Coast, where buyers for Asian recyclers are most active. The lowest values are in eastern Canada, which has several nearby primary smelters. If located in another country, export taxes may also reduce the value of the scrap (Foster, 2005).
- *The location of markets for recycled aluminum*: The customers for recycled aluminum can be a long way from the place where it is produced. Most

TABLE 4.2

Price for Segregated Low-Copper Clips (Taboo), $US/lb

City	Price ($US/lb)
Atlanta	0.6464
Boston	0.6114
Buffalo	0.6414
Chicago	0.6264
Cincinnati	0.6214
Cleveland	0.6064
Detroit	0.6314
Houston	0.5514
Los Angeles	0.6814
New York	0.6264
Philadelphia	0.6364
Pittsburgh	0.6264
San Francisco	0.6564
St. Louis	0.6664
Montreal	0.5273
Toronto	0.5273

secondary aluminum producers locate their plants near their customers; those that cannot will receive less for their product, which means that they pay less for raw material.

- *The cost of alloying elements*: Aluminum alloys generally contain one or more of five common alloying elements: copper, magnesium, manganese, silicon, and zinc. If the cost of any of these elements surges, scrap containing that element can become a desirable alternative to pure copper, magnesium, etc., as a means of manufacturing that alloy. The costs of copper and silicon can make the price of scrap containing these elements higher than that of primary metal (Blomberg and Hellner, 2000; Gaustad et al., 2007; Rombach, 2006).
- *Government policy* (see Chapter 11): Taxes and restrictions on imports or exports of aluminum affect scrap prices, depending on the type of government action (Foster, 2005). One government policy that has had a particular impact has been moves to reduce the carbon *footprint* of manufacturing industries, that is, the amount of CO_2 released to the environment during the production process. Because the production of recycled aluminum has such a smaller carbon footprint than that of primary metal (see Chapter 1), manufacturers wishing to lower their own carbon impact can do so by using recycled metal (Das, 2012; Novelis, 2012). This increases the demand for scrap and raises its price.

Table 4.3, obtained from Gesing et al. (2010), compares the relative value of prices for different grades of scrap with that of primary metal in spring 2010. The most valuable scrap grade is alloy 6061 truck wheels, which consist of a single alloy, have no coating, and so can be directly remelted to the same alloy. The second most

TABLE 4.3

Relative Price for Aluminum Scrap Grades in Spring 2010

Scrap Type	Origin	ISRI Code	Scrap Price (as % of Primary)
Mixed-alloy, high-grade turnings	New	Telic	15.3
Sheet	Old	Taint	21.4
Cast	Old	Tense	24.7
Siding	Old	Tale	28.4
UBC (cans)	Old	Talc	32.1
Mixed low-copper clips	Old	Tabor	38.1
Siding	New		44.2
Mixed low-copper clips	New	Taboo	58.6
Litho sheet	Old	Tablet	65.1
6063 extrusion (+6061)	New	Toto	70.7
Alloy 356 wheels	Old	Troma	78.6
Alloys 1100 + 3003	New	Tabloid	81.4
Alloy 5052	New	Tooth	82.8
Alloy EC scrap	New	Talon	85.1
Alloy 6061	New	Tutu	85.6
Alloy 6063	New	Tata	89.3
Alloy 6061 truck wheels	Old		89.8
1-1-3 sow	Old	Throb	42.3

Source: Gesing, A. et al., Advanced industrial technologies for aluminium scrap sorting, presented at *Aluminium-21/Recycling*, St. Petersburg, Russia, 2010.

valuable grade is new alloy 6061 scrap, which has the advantage of being new, but may not have the consistent shape and size of old wheels. The lowest-valued scrap is mixed-alloy, high-grade turnings (Figure 4.4), which cannot be directly remelted, contain oil or lubricant, have a high surface/volume ratio, and may have more copper than most aluminum alloys can tolerate. The sorting technologies described in the next chapter will not be useful for upgrading this material, either. As a result, its usefulness is limited. In general, the lowest-valued scrap grades tend to be old rather than new, mixed rather than single alloy, and coated or oily rather than bare. The exception is the 1-1-3 secondary ingot, produced by sweat melting (see Chapter 5). Its value is lowered by the iron content, which is incompatible with many aluminum alloy specifications (see Chapter 2).

Figure 4.5 shows another characteristic of aluminum scrap prices, volatility (Brath, 2012). The graph compares recent prices of scrap grades Taint and Tabor (bottom curve), secondary aluminum ingot (middle curve, described later in this chapter), and primary metal (top curve). These prices are graphed relative to the price of primary metal at the beginning of 2009. The up-and-down behavior of Taint/Tabor is a characteristic of all types of scrap and a challenge for recyclers trying to manage and if possible reduce costs (NADCA, 2008). The reason for the volatility is that the supply of scrap is not strongly affected by price. The amount of

FIGURE 4.4 Mixed borings and turnings (ISRI grade Telic). (Photo courtesy of Ray Peterson, Aleris International, Beachwood, OH.)

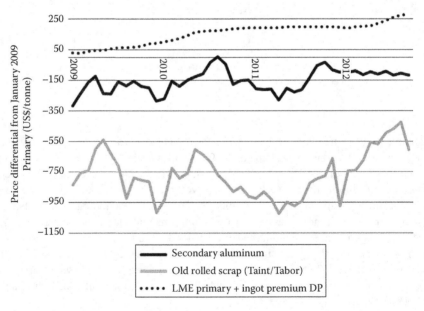

FIGURE 4.5 Prices of primary aluminum metal, secondary ingot, and scrap grade Taint/Tabor. (From Brath, C.F., Market development in primary and recycled aluminium, presented at *Metal Bulletin 20th International Recycled Aluminium Conference*, Salzburg, Austria, November 20, 2012. With permission.)

new scrap is determined by the activities of aluminum processors, not scrap consumers (Blomberg and Hellner, 2000), and actions that would increase the supply of old scrap are often the result of government action, not economic. Blomberg and Söderholm (2009) estimate that the own-price elasticity of secondary aluminum supply is 0.21, meaning that a 10% increase in price is likely to increase the supply by only 2.1%. As a result, increases in demand for scrap cause a significant increase in price. Another consequence of the low own-price elasticity is that increasing the supply of aluminum scrap is less likely to be accomplished by providing price incentives than by improving recycling technology (Xiarchos, 2005).

Prices for aluminum scrap are published by numerous sources. Many provide *spot prices,* which are the prices paid for scrap for immediate (2-day) delivery. The AMM (*American Metal Market*) prices shown in Table 4.2 are an example of these (AMM, 2012); numerous web-based services have also appeared in the past several years (Logue, 2012a). The prices quoted by these sources are a combination of auction results, conversations with scrap brokers and buyers, and surveys of participants. Spot prices for scrap show the greatest volatility. Still other listings are based on an average of previous spot prices; the industry publication *Platts Metals Week* (now *Platts Metals Daily*; see Table 4.1) uses the average for a prior month for a given scrap grade to set the value for the next month (NADCA, 2008).

The London Metals Exchange (LME) does not publish scrap prices, but it does report the price of alloy A380.1, a casting alloy produced from scrap. The formula used by the LME uses the average price of this alloy from two months before to set the price for the current month (NADCA, 2008). This matters because contracts for the purchase and sale of scrap often use the LME price for A380.1 as a reference (Anton, 2006).

SCRAP PURCHASING

A recycler can essentially purchase scrap one of two ways:

1. On the spot market, either through auction or through a broker (see Chapter 3)
2. By contract with a scrap generator or broker, using a price or price formula determined by negotiation

A recycler who obtains scrap via the first method is engaging in *commodity risk,* the risk that aluminum prices will fall during the remelting or refining process. This reduces or eliminates the profit margin obtained by recycling the metal. However, an increase in aluminum alloy prices will increase the margin, making the operation more profitable. A second risk of spot market purchases (and perhaps more serious) is that of quantity; some scrap grades may not be widely available, especially at certain times of the year. As a result, most recyclers avoid this approach.

Recyclers who buy scrap by contract reduce their quantity risk. However, they can still assume a commodity risk, if the negotiated purchase is a function of a spot price (for aluminum scrap or aluminum). A particular concern is that of buying scrap at a price based on one index and selling castings or billet at a price based on another (NADCA, 2008). Recyclers often deal with this concern by *hedging* their scrap purchases. Hedging describes several techniques used to lock scrap prices into place, minimizing the risk of sudden changes in price that could result in losses to the recycler. It is not without risks of its own (Logue, 2012b). There are several hedging techniques available; the most significant for aluminum recyclers will be discussed here.

Physical hedging is the practice of buying large quantities of scrap at a low price and storing it on site (Logue, 2012b). If the recycled product generated from the scrap can be sold the same day, the profits can be *locked in*. If not, the recycler is gambling that the price of secondary aluminum will be high enough when the metal is sold to generate a profit. Physical hedging requires sufficient understanding of the market to know when a scrap price is low enough to justify buying. It also requires substantial storage space for the purchased scrap, which is costly and ties capital up in inventory. As a result, its use is limited to larger recyclers.

If a recycler does not have the storage space for large quantities of scrap or the desired quantities of a particular grade of scrap are not available at one time, *derivatives hedging* can be used to reduce risk. Derivatives hedging is done by purchasing scrap through a *futures contract,* a commitment by the recycler to purchase a specified quantity of scrap at a set time in the future (LME, 2012; NADCA, 2008). The price of the scrap is either a set value (*fixed price contracting*) or more likely a function of the market price of a more stable commodity (*formula pricing*). Although a futures contract guarantees physical delivery of the contracted amount of the scrap at the specified time, in practice this rarely happens. The recycler usually finds a cheaper source of the desired scrap instead. When this happens, the futures contract is sold at the original metal price. This limits the profits made by the recycler from the less expensive scrap, but the guaranteed price also minimizes losses should the purchase of more expensive scrap be required instead.

Finding a basis for calculating the purchase price of aluminum scrap in a futures contract has been a challenge. As previously mentioned, the price of primary metal does not always correspond to the price of scrap (Anton, 2006), and published scrap prices are too volatile to be used for this purpose. The prices of secondary alloys (alloys produced from scrap) should be more relevant for this purpose. However, a single secondary alloy can have several compositions, as will be seen later in this chapter, and so prices published for that alloy in different sources may not be for the same material. As a result, the type of contract available for derivatives hedging differs from continent to continent.

In the early 1990s, the LME attempted to solve this problem by introducing futures contracts based on the price of alloy A380.1 (Kiser, 2002). *Aluminum alloy* contracts were offered for Europe, North America, and Japan. The contracts for Europe were initially successful, but the A380.1 used in North America is different from that used in Europe (Carey, 2013), and the price is different as well. As a result, American secondary aluminum producers and consumers (in particular automakers) sought a better hedging tool. This resulted in the introduction by the LME of the North American Special Aluminum Alloy Contract (NASAAC) in 2002 (Anton, 2006; Stundza, 2002). NASAAC applied only to a North American composition of A380.1, to be supplied from US warehouses. This made it compatible with US accounting rules and thus a more usable tool.

The LME contracts have been controversial in both Europe and North America. Scrap dealers in particular have been opposed to it (Carey, 2013), noting that the difference between the price of LME A380.1 and that from other sources (in particular

Platts) is increasingly large and volatile (McBeth, 2012). Futures contracts imply the ability to obtain metal from LME-approved warehouses at the specified time, but waiting periods of several months have become increasingly common (Burns, 2012). As a result, secondary aluminum customers have been forced to make adjustments to NASAAC prices, and producers of secondary aluminum have begun to look for other hedging mechanisms instead (McBeth and Baltic, 2012).

COST OF PROCESSING

The cost of processing scrap is difficult to state precisely, because there is no *typical* processing flowsheet. Some recyclers start with low-cost old scrap and produce secondary alloys for casting applications; others start with good-quality new scrap and remelt directly to the same alloy. Analysis of a unit operation as simple as remelting scrap depends on the fuel value (if any) of coatings on the scrap, the melt loss that occurs during the process, and the final temperature (casting alloys have a lower melting point than wrought alloys). However, some analyses have been produced.

For most producers of recycled aluminum, the cost of scrap represents about 80%–85% of total costs (Kozhanov et al., 2010; Luo, 2009). As a result, the most effective means of reducing costs is minimizing the cost of scrap. Hedging is one possibility; finding ways to replace new scrap with old is another. The development of technology to more effectively sort old scrap (see next chapter) and to remove alloying elements from molten aluminum (see Chapter 12) is important to the industry for this reason.

Figure 4.6 illustrates the breakdown by Gluns and Schemberg (2005) of processing costs for a secondary aluminum smelter burning natural gas in air. The largest components are the cost of disposing process wastes and labor costs; as a result, reducing these has the highest priority. Process wastes include salt slag and dross, and ways to reduce the amount of each to be disposed of are described in Chapter 13. Unfortunately, the generation of these wastes increases when lower-grade scrap is melted, which puts this cost-reduction effort at cross-purposes with that of reducing

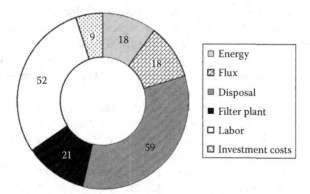

FIGURE 4.6 Costs of processing aluminum scrap, €/tonne. (From Gluns, L. and Schemberg, S., Advantages of oxy-fuel burner systems for aluminium recycling, https://www. airproducts.com/~/media/Files/PDF/industries/metals-advantages-oxy-fuel-burner-systems-aluminium-recycling.pdf, 2005. With permission.)

scrap costs. The reduction of labor costs requires the use of greater automation in furnace operations, described in Chapter 9. This too is more difficult to accomplish when lower-grade scrap is used. Minimizing the use of salt would reduce flux as well as disposal costs. The recent surge in natural gas production has reduced fuel costs; better furnace design (see Chapters 7 and 8) will reduce it further.

RECYCLING PRODUCTS AND THEIR PRICE

There are two primary types of aluminum recycling facility. The first is a *remelter*. Remelters engage in *direct recycling,* producing a specific alloy from scrap primarily consisting of that alloy. As discussed in Chapter 11, some recycling facilities are now located adjacent to manufacturing facilities and effectively engage in *toll processing* of their scrap. In such cases, the price of the alloy being produced is unimportant; the recycler is essentially charging a fee for processing the material and returning it in usable form to the original source of the scrap (AlFed, 2011). In other places, the supply of scrap consisting of one alloy is sufficient in some markets to justify seg-regating it and remelting it separately. The scrap grades listed in Figure 4.1 include several that consist of just one alloy. If the cost of remelting and casting is less than the price difference between that grade of scrap and the market value of that alloy, the recycler can make a profit.

Secondary smelters, sometimes known as *refiners,* operate differently. They pro-duce alloys for sale to the aluminum industry at large, from a variety of scrap sources that can change depending on the price and availability of a given type of scrap. The standards that their products meet are common to all customers, and prices are pub-licly listed. A description of these products will describe what influences their prices and why they are more commonly produced than other possible alloys.

Table 4.4 lists the chemical composition of seven compositions of popular alu-minum casting alloy "380." This alloy features high copper and silicon levels and optionally up to 3.0% zinc. The first distinction among the seven types is the product type. Alloys with "0.0" at the end (380.0, A380.0, and B380.0) are alloys that have been die cast; castings are often sold by the pound or kilogram, with a markup for processing costs. The "0.0" alloys have the highest permissible levels for impuri-ties such as iron, nickel, and tin. The "0.1" and "0.2" alloys are ingots cast by the smelter; sows also fall under this category. Typically, "0.1" alloys are those primar-ily produced from scrap by secondary smelters, and "0.2" alloys are produced using primary metal. Because primary metal has fewer impurities, the acceptable limits for iron, nickel, and tin are lower in "0.2" alloys. The impurity limits in A380.2 are especially noteworthy; they include specific limits of 0.05% on nonlisted individual impurity elements, as well as an overall limit of 0.15%. As the number of alloying elements used in the aluminum industry grows, keeping unusual impurities out of scrap becomes an increasing challenge (Carey, 2013). A380.2 and B380.2 also have higher silicon levels than their "0.1" counterparts; lower allowable levels of iron and nickel ensure that this much silicon can be added without the risk of forming an intermetallic precipitate. The difference between A380.1 and the other alloys is the higher level of zinc (die casters producing 380.0 from 380.2 ingot have to add the zinc themselves). The higher level of zinc gives secondary smelters greater choice

TABLE 4.4

Composition of Different Types of Casting Alloy "380"

Alloy	Product	Cu	Mg	Si	Fe	Mn	Ni	Zn	Sn	Others (Each)	Others (Total)
A380.0	Die cast	3.0–4.0	0.1	7.5–9.5	1.3	0.5	0.5	3.0	0.35		0.50
A380.1	Ingot	3.0–4.0	0.1	7.5–9.5	1.0	0.5	0.5	2.9	0.35		0.50
A380.2	Ingot	3.0–4.0	0.1	9.5–11.5	0.6	0.5	0.1	0.1	0.10	0.05	0.15
380.0	Die cast	3.0–4.0	0.1	7.5–9.5	2.0	0.5	0.5	3.0	0.35		0.50
380.2	Ingot	3.0–4.0	0.1	9.5–11.5	0.7–1.1	0.5	0.1	0.1	0.10		0.20
B380.0	Die cast	3.0–4.0	0.1	7.5–9.5	1.3	0.5	0.5	1.0	0.35		0.50
B380.1	Ingot	3.0–4.0	0.1	7.5–9.5	1.0	0.5	0.5	0.9	0.35		0.50
LME NASAAC 380.1	Ingot	3.0–4.0	0.1	8.0–9.5	1.0	0.5	0.5	2.9	0.35		

of scrap grades that they can use in their mix, making A380.1 more *recycle friendly* than B380.1 (Das et al., 2008). Table 4.4 also includes the composition of the *special* North American grades of A380.1 specified for NASAAC by LME. Similar composition *families* exist for other common grades of secondary alloy as well, each with its own price.

Recycled aluminum can be delivered to customers as molten metal (see Chapter 8) or as a solid product. The three most common solid forms are ingot, sow, and T-bar (Grandfield, 2009). Ingots range in size up to 23 kg (50 lb), sows are typically 500–1000 kg (Figure 4.7), and T-bars tend to be less than 25 kg. Customers tend to prefer ingot, in part for safety reasons (Carey, 2013); sows and T-bars can trap water during solidification, creating an explosion hazard during remelting (see Chapter 14). Grandfield (2009) discusses this in greater detail.

Table 4.5, also derived from Gesing et al. (2010), lists the price of several common secondary alloys as a percentage of the price of primary aluminum in spring 2010. Most of the prices are higher than that of primary metal, largely due to the high percentage of expensive alloying agents such as copper and silicon. Comparing the price of 356.1 (produced from scrap) and 356.2 (produced from primary) shows the value to customers of the lower impurity content in "0.2" type alloys. Also included are the prices of two wrought alloys produced by remelters, 6061 and 6063. Alloy 6061 has higher alloy content; 6063 has tighter impurity restrictions. The price difference between "A380.1" and "LME 380" is one of the reasons why the NASAAC is poorly regarded by some recyclers (Carey, 2013; McBeth, 2012).

Figure 4.8, from the presentation by Borner (2012), illustrates the spread during 2010–2011 between the US price of A380.1 reported by Platts and the market price of *auto shreds* (i.e., ISRI grade Twitch). Two curves are presented. The top curve shows the spread between the two listed prices, and the bottom shows the spread when the effect of melt loss during smelting is taken into account. The decline at the

FIGURE 4.7 Low-profile secondary aluminum sow. (Photo courtesy of Ray Peterson, Aleris International, Beachwood, OH.)

TABLE 4.5

Prices for Secondary Aluminum Products in Spring 2010

Product	Price (as % of Primary)
LME 380	83.3
A380.1	91.6
319.1	100.9
A413.1	110.2
A360.1	110.2
384.1 (384 + 3% Zn)	114.0
356.1	117.2
332.1	123.7
B380.1 (380 + 1% Zn)	124.7
6061 extrusion billet	125.1
443.1	130.7
6063 extrusion billet	138.1
355.2	141.4
356.2	147.0

Source: Gesing, A. et al., Advanced industrial technologies for aluminium scrap
sorting, presented at *Aluminium-21/Recycling*, St. Petersburg, Russia,
2010.

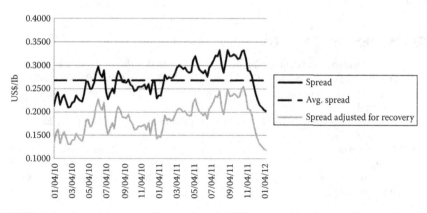

FIGURE 4.8   Spread between Platts price of A380.1 secondary ingot and upgraded auto
shred (ISRI Twitch). (From Borner, G., Secondary scrap demand for aluminum, presented
at *Platts Metals Week Aluminum Symposium*, Fort Lauderdale, FL, http://www.platts.com/
IM.Platts.Content/ProductsServices/ConferenceandEvents/2012/gc203/presentations/Gary_
Borner.pdf, 2012.)

end of 2011 reflects a decline in the price of primary aluminum. Subtracting the cost
of processing from this spread determines the profitability of the smelter. Figure 4.9
shows the result when the NASAAC price is used instead. The spread declined to
nearly zero at the end of 2011, before subtracting the processing cost. The volatility in
these curves explains why a hedging device is needed; the spread shown in Figure 4.9

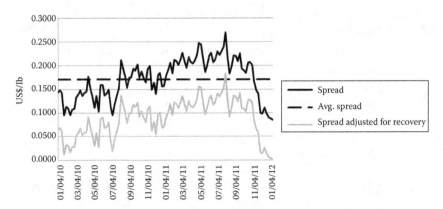

FIGURE 4.9   Spread between NASAAC price of A380.1 secondary ingot and Platts price of upgraded auto shred (ISRI Twitch). (From Borner, G., Secondary scrap demand for aluminum, presented at *Platts Metals Week Aluminum Symposium*, Fort Lauderdale, FL, http://www.platts.com/IM.Platts.Content/ProductsServices/ConferenceandEvents/2012/gc203/presentations/Gary_Borner.pdf, 2012.)

shows why the NASAAC may not be the best choice for this device. The rest of this book will highlight means of reducing the processing cost for aluminum scrap; however, the importance of improving the financial structure of the industry should not be forgotten.

RECOMMENDED READING

Borner, G., Secondary scrap demand for aluminum, presented at *Platts Metals Week Aluminum Symposium*, Fort Lauderdale, FL, January 17, 2012. http://www.platts.com/IM.Platts.Content/ProductsServices/ConferenceandEvents/2012/gc203/presentations/Gary_Borner.pdf.
Carey, R., NASAAC—An analysis of a capitalist system gone wrong, *Die Cast. Eng.*, 56(1), 36, January 2013.
NADCA (North American Die Casting Association), *White Paper: Effective Risk Management Strategies Reduce Commodity Volatility for Die Casters*, North American Die Casting Association, Wheeling, IL, April 18, 2008. http://www.tvtdiecast.com/documents/whitepaper11.pdf.

REFERENCES

AlFed (Aluminium Federation), *Aluminium Wrought Remelt*, Aluminium Federation, West Midlands, U.K., September 6, 2011. http://www.alfed.org.uk/downloads/documents/LBUAISFRIK_7_aluminium_wrought_remelt_1_.pdf.
AMM, AMM monthly averages: Nonferrous scrap prices, *American Metal Market,* June 8, 2012.
Anton, K.J., Beyond hedging: Using futures as a management tool, *Fut. Ind.*, 16(5), 21, September/October 2006. http://www.aluminum.org/Images/AluminumNow/0902/0902feature2.htm.
Blomberg, J. and Hellner, S., Short-run demand and supply elasticities in the West European market for secondary aluminium, *Res. Policy*, 26, 39, 2000.

Blomberg, J. and Söderholm, P., The economics of secondary aluminium supply: An econometric analysis based on European data, *Res. Conserv. Recycl.*, 53, 455, 2009.

Borner, G., Secondary scrap demand for aluminum, presented at *Platts Metals Week Aluminum Symposium*, Fort Lauderdale, FL, January 17, 2012. http://www.platts.com/IM.Platts.Content/ProductsServices/ConferenceandEvents/2012/gc203/presentations/Gary_Borner.pdf.

Brath, C.F., Market development in primary and recycled aluminium, presented at *Metal Bulletin 20th International Recycled Aluminium Conference*, Salzburg, Austria, November 20, 2012.

Burns, S., EU anti-competition watchdog reviewing LME warehouse load-out rates, MetalMiner, December 6, 2012. http://agmetalminer.com/2012/12/06/eu-anti-competition-watchdog-reviewing-lme-warehouse-load-out-rates/.

Carey, R., NASAAC—An analysis of a capitalist system gone wrong, *Die Cast. Eng.*, 56(1), 36, January 2013.

Das, S., Achieving carbon neutrality in the global aluminum industry, *JOM*, 64(2), 28, February 2012.

Das, S.K., Green, J.A.S., and Kaufman, J.G., Application of the recycling indices to the identification of recycle-friendly aluminum alloys, November 6, 2008. http://www.phinix.net/services/Recycling/RecycleIndex.pdf.

Foster, K., China export tax said damaging aluminum recycling industry, *Am. Metal Market*, March 8, 2005.

Gaustad, G., Li, P., and Kirchain, R., Modeling methods for managing raw material compositional uncertainty in alloy production, *Res. Conserv. Recycl.*, 52, 180, 2007.

Gesing, A., Erdmann, T., and Gesing, M.A., Advanced industrial technologies for aluminium scrap sorting, presented at *Aluminium-21/Recycling*, St. Petersburg, Russia, 2010.

Gluns, L. and Schemberg, S., Advantages of oxy-fuel burner systems for aluminium recycling, February 14, 2005. https://www.airproducts.com/~/media/Files/PDF/industries/metals-advantages-oxy-fuel-burner-systems-aluminium-recycling.pdf.

Grandfield, J.F., Remelt ingot production technology, in *Light Metals 2009*, Bearne, G., Ed., TMS-AIME, Warrendale, PA, 2009, p. 851.

ISRI (Institute of Scrap Recycling Industries), *Scrap Specifications Circular 2012*, October 8, 2012. http://www.isri.org/iMIS15_Prod/ISRI/_Program_and_Services/Commodities/Scrap_Specifications/ISRI/_Program_and_Services/Scrap_Specifications_Circular.aspx?hkey=5c76eb15-ec00-480e-b57f-e56ce1ccfab5.

Kiser, K., Birth of an aluminum contract, *Scrap*, 59(3), 26, May/June, 2002.

Kozhanov, V.A., Popov, V.A., and Mushik, E.E., Ways to increase competitiveness and efficiency of metal goods production from scrap and nonferrous waste, *Metall. Min. Ind.*, 2, 152, 2010.

LME (London Metal Exchange), *Education: Training Resources: Videos*, November 29, 2012. http://www.lme.com/education/training-resources/videos/.

Logue, A.C., Finding the right price, *Scrap*, 69(5), 76, September/October 2012a.

Logue, A.C., Managing risk, *Scrap*, 69(6), 89, November/December 2012b.

Luo, Z., The impact of emissions trading on the secondary aluminium industry of the EU, *Aluminium*, 85(3), 21, March 2009.

McBeth, K., Auto OEMs should change pricing base for alloys: Moore, *Platts Metals Daily*, 1(4), 3, October 15, 2012.

McBeth, K. and Baltic, S., Aleris to discontinue use of NASAAC index in aluminum pricing, *Platts RSS Feed*, October 9, 2012. http://www.platts.com/RSSFeedDetailedNews/RSSFeed/Metals/6684016.

NADCA (North American Die Casting Association), *White Paper: Effective Risk Management Strategies Reduce Commodity Volatility for Die Casters*, North American Die Casting Association, Wheeling, IL, April 18, 2008. http://www.tvtdiecast.com/documents/whitepaper11.pdf.

Novelis, *Novelis Sustainability Report 2012: A Whole Life Cycle Approach*, August 27, 2012. http://286223ab820fcdd418c1-a70764a0e89d03092c3be35911f200c1.r41.cf1.rackcdn. com/Novelis_FY12_Sustainability_Report.pdf.

Platts Metals Week, *Methodology and Specifications Guide*, May 2, 2012. http://www.platts. com/IM.Platts.Content/methodologyreferences/methodologyspecs/metals.pdf.

Rombach, G., Limits of metal recycling, in *Sustainable Metals Management*, von Gleich, A., Ayres, R.U., and Gößling-Reisemann, S., Eds., Springer, Dordrecht, the Netherlands, 2006, p. 312.

Stulgis, G., Profitability of recycled aluminium industry in Europe, presented at *Metal Bulletin 20th International Recycled Aluminium Conference*, Salzburg, Austria, November 20, 2012.

Stundza, T., Buyers eye use of new secondary contract, *Purchasing*, 131(6), 16B1, 2002.

VDM (Verband Deutscher Metallhändler e.V.), *Usancen und Klassifizierungen des Metallhandels*, November 27, 2011. http://www.metallhandel-online.de/downloads/usancen_2012.pdf.

Xiarchos, I.M., Time-varying ratios of primary and scrap metal prices: Importance of inventories, Chapter 5 in Commodity Modeling and Pricing, Schaeffer, P.V., Ed., John Wiley & Sons, New York, 2008.

5 Beneficiation Technology

After mining, the next step in traditional metal extraction is minerals processing, in which the valuable minerals are liberated from the gangue material in the ore. Following this, the minerals are sorted into a separate stream, using differences between their physical properties and those of the gangue. In some cases, chemical or thermal processes may be used to remove impurities from the concentrated minerals. As a final step, the concentrated minerals may be agglomerated to make them easier to ship and easier to handle in subsequent smelting processes.

In Chapter 3, the mining (collecting) of scrap aluminum was described. Over the next two chapters, the equivalent mineral processing (*beneficiation* or *upgrading*) of this ore will be introduced. In this chapter, the techniques and equipment used to separate scrap aluminum from contaminants will be introduced. This will be followed by a description in Chapter 6 of how this technology is put into practice. As with the processing of traditional minerals, the beneficiation of scrap can be divided into four types of unit process. These include comminution, sorting, thermal processing, and agglomeration.

COMMINUTION (NIJKERK AND DALMIJN, 1998)

Aluminum scrap comes in a wide variety of size and conditions. Pieces can be as small as the aluminum wires used in electronic equipment or as large as a jet airliner. Scrap is sometimes *liberated*, free of any attached pieces of other materials. However, it is often found bolted, welded, or otherwise attached to other parts. To be remelted, the scrap will have to be separated from these other materials. The dismantling of automobiles and other large assemblies described in Chapter 3 accomplishes this to some degree. However, in many cases, the value of the scrap recovered by dismantling is insufficient to justify the expense of this labor-intensive process.

Many sorting processes require scrap pieces of a given size to be effective. In addition, furnaces have maximum sizes of scrap that can be successfully fed. Consistent particle size is also essential to the proper functioning of some sorting devices. Some scrap already has the proper size for automated sorting, but most of it does not. *Comminution* is the process by which oversize material is reduced to the proper size for further processing. The comminution of naturally occurring minerals is usually done by crushing it into smaller rocks. However, metal scrap cannot be turned into smaller separate pieces in this fashion. Instead, it must be torn apart. Numerous devices have been invented for tearing apart metal scrap. Here, they will be separated into three categories—*shears, impact shredders,* and *rotary shredders.*

Most of the shears in operation are either *alligator* or *guillotine* shears. Figure 5.1 shows a typical alligator shear (Fowler, 2003), so named because the hinged cutting motion is similar to that of an alligator. Only the top portion of the jaw moves

FIGURE 5.1 An alligator shear. (From Ohio Baler Co., http://www.ohiobaler.com/alligatorshear.htm. With permission.)

during operation. The cutting blade is made of hardened tool steel and can be resharpened when it gets dull. Sizes and power ratings vary, according to the size and type of material to be cut. Most alligator shears use electric power, although gasoline and diesel engines are available.

Alligator shears have been available since the early 1900s. Major changes in the basic design include the following:

- The use of hydraulic rather than mechanical flywheel drives. Hydraulic drives allow the machine to be stopped immediately to prevent damage to the machine or injury to the operator, a feature not available in flywheel-driven shears.
- The addition of a hold-down clamp to keep scrap pieces still while being cut.
- The addition of guards around the machine to prevent injury to the operators. Alligator shears were notorious safety hazards before the introduction of these guards, which are now mandatory in most countries.

One of the advantages of the alligator shear is its versatility (A-Ward, 2005). This is particularly important in the processing of aluminum scrap, which takes a variety of forms. Figure 5.2 shows a crane-mounted shear used for recovering scrap that cannot be brought to a processing plant. Crane-mounted shears remove aluminum and other scrap metals from aircrafts, ships, rail cars, and other large objects. Another special type of alligator shear is the nibbling or metal cleaning shear used for detaching small metal attachments such as bolts or collars. Metal-cleaning shears are especially popular for separating steel parts from cable scrap (American Recycler, 2003).

FIGURE 5.2   A crane-mounted shear. (From Harding Metals, Inc., http://hardingmetals. com/company.htm. With permission.)

FIGURE 5.3   A guillotine shear. (From Gershow Recycling, http://www.gershowrecycling. com. With permission.)

Figure 5.3 shows a *guillotine* shear, which operates the same way as the well-known device used for executing persons (Pretz and Julius, 2011). Scrap is placed underneath a cutting blade, which drops down onto the piece with the help of vertical guide rails. Since scrap is more difficult to chop than human heads, hydraulic power is used to accelerate the blade rather than relying on gravity alone. Guillotine shears are more productive than alligator shears and can be built with higher power ratings. As a result, they have replaced alligator shears in many applications, particularly for cutting steel scrap. However, they are less important in the processing of scrap aluminum. A recent innovation is a *swing beam* shear (Marshall, 2012), which pivots the blade about a fixed point. This allows it to cut thinner material than a guillotine shear.

Describing all the types of shredder would fill a book by itself. The discussion provided here summarizes the more important types. For a more complete description, the reader is encouraged to see other sources, in particular Nijkerk and Dalmijn (1998).

Figure 5.4 shows a cutaway view of an older model of *hammer mill,* the most popular type of impact shredder. These were first used for shredding automobiles in the late 1950s (Manouchehri, 2007) and have since become the predominant means for processing auto hulks. They are also used for processing white goods such as refrigerators and ovens and have been adapted to process smaller items as well. The model shown here is used for shredding auto hulks but is similar in function to those used for other purposes.

The description of a hammer mill starts with the rotor assembly (Manouchehri, 2007). The rotor is several feet long and has a series of *spider* assemblies attached to it. The assemblies have wear-resistant steel caps; attached to each cap is a hammer made of ductile iron or manganese steel. The hammers weigh 50–300 kg depending

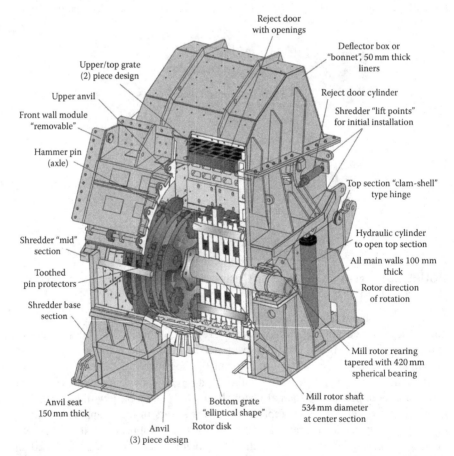

FIGURE 5.4 Exploded view of hammer-mill shredder. (From Metso Texas Shredder. With permission.)

on the capacity of the hammer mill and are allowed to swing freely as the rotor turns. The rotor turns at 500–600 rpm, requiring up to 7000 hp. Faster rotation speed uses more energy but reduces the time required to reduce input scrap to a given size. As a result, for a set product size, the specific energy consumption of shredding decreases with increasing rotor speed. Horizontal rotor arrangements such as that shown here are preferred for large-scale shredding of metallic scrap over vertical hammer mills. While AC motors are the predominant choice for larger hammer mills, DC motors have become more popular for small- and medium-size units.

Figure 5.5 shows a second view of a hammer mill, in cross section. Scrap fed to a hammer mill is first passed through a flattener or compressor (part 2 in the figure). This device (1) increases the density of the feed, which improves the operation of the shredder (Manouchehri, 2007), and (2) controls the rate at which the scrap is fed to the machine, ensuring steady-state operation. When the scrap enters the main chamber, a breaker bar absorbs the shock. The swinging hammers then engage the scrap, pounding it repeatedly against the anvil or cutter bar (part 3). (Remember the mass of the hammers and the rotor speed.) This action tears the scrap apart, turning it into fist-size chunks that go through the holes in the manganese steel grate (part 4).

Although shredders are extremely powerful machines, occasionally a scrap piece will come through that is too thick or massive to be shredded. The reject door (part 9 in Figure 5.5) allows these pieces to be removed from the shredder before they can damage the machine. Other oversize pieces simply go around and are eventually caught against a baffle plate (part 7). These have a similar function as the anvil and give the hammers another chance to finish the job. Manouchehri (2007) suggests that most of the size reduction in hammer mills actually occurs at this location. A second grate is provided at the top (part 5) to allow shredded material and fluff a second chance to leave the shredding chamber. This reduces the load of returned material and improves shredder capacity.

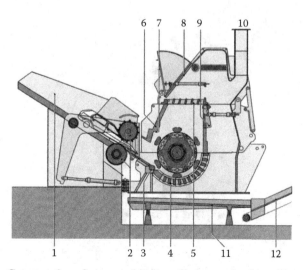

FIGURE 5.5 Cross section of a top-and-bottom discharge shredder. (From Metso Texas Shredder. With permission.)

All shredders generate dust and fine particles during operation. These would be an environmental and workplace hygiene hazard if released. Furthermore, shredder operation generates heat. If the dust contains organic vapors from vaporizing grease or oil in the scrap, this can result in a significant explosion hazard. As a result, all dry shredders have a dust collection system (part 10 in Figure 5.5), operating under negative pressure to draw the dust off (Gesing, 2006).

One option for eliminating the dust from shredding operations is to spray water into the shredder (Gesing, 2006; Manouchehri, 2007). *Moist* shredding adds the water as a mist and minimizes the amount added; *wet* shredding floods the shredder and generates a slurry containing the shredded material. This eliminates the need for dust collection and minimizes the risk of explosion. However, there are disadvantages to moist or wet shredding as well. The first is that it introduces the need for a water-treatment system to recycle the runoff; any soluble material in the feed (e.g., road salt) is a potential water pollution problem. A second is the need to recover what would have been the dust somewhere else. The potential for the dust to contaminate the metal streams recovered from the shredder is a product quality concern. The loss of metal fines in the runoff can also become an issue. The heat generated by shredding turns added water into steam, which makes monitoring difficult; some moist or wet shredders use an infrared camera to deal with this (Fowler, 2007a). As a result, moist shredding has become more popular than wet shredding. The amount of water needed can be reduced by adding foam (Midwest, 2012).

Figure 5.6 shows the cross section of an *impact crusher*, also used to shred large and heavy scrap. The impact crusher is similar to a hammer mill but uses hammers fixed to the spider assembly rather than being allowed to swing freely. A series of breaking surfaces replaces the single anvil in the hammer mill. The impact crusher is used more frequently to crush minerals than metallic scrap. For scrap processing, it is occasionally used to "*recrush*" metal than has been processed in a hammer

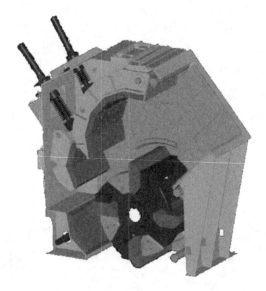

FIGURE 5.6 View of an impact crusher. (From Roc–Impact s.a.r.l., (2013) http://www. crusher-rocimpact.com/crushers/impact-crusher.php. With permission.)

FIGURE 5.7 Double-rotor rotary shear. (From Metso Minerals, http://www.metsominerals. com/inetMinerals/MaTobox7.nsf/DocsByID/F9790232BD1B5BF442256B440047DC5B/ $File/1243_RO_EN.pdf. With permission.)

mill (Lindemann, 2009). This reduces the required degree of reduction in the hammer mill and increases productivity.

Hammer mills were designed to shred large pieces of ferrous scrap, such as those in automobiles. As a result, they use tremendous amounts of energy, generate large quantities of fines from more easily shredded materials, and are very noisy. For shredding lighter and weaker materials, more cost-effective machines are available. Figure 5.7 shows a double-rotor *rotary shredder* (also known as a *rotary shear*). These shredders are more widely used for processing paper and plastic (Fowler, 2011) but process light metal scrap as well, particularly foil and UBCs. They are also the primary device used for wire chopping to recover metal from cable scrap (Sullivan, 1985). As with hammer mills, many designs are available.

The rotors in a rotary shredder turn in opposite directions. Each has several knife blades attached, such as those in Figure 5.8. Each knife blade includes one or more

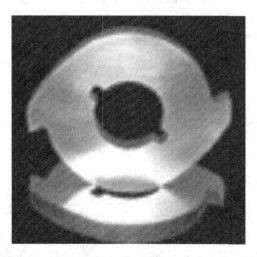

FIGURE 5.8 Knives for a rotary shredder. (From Komar Industries, Inc. (2013) http://www. komarindustries.com/equipment/electro-shear-shredder.php. With permission.)

hooks. The hooks grab pieces of scrap and pull them down into the space between the rotors, where the knives slice them into smaller shreds. The shreds fall either directly onto a conveyor that takes them away or onto a screen that makes the return of oversized pieces to the shredder possible. The size of the shreds is determined by the shredder design; increasing the number of rotors decreases the size, as does reducing the rotor diameter. Rotary shredders use electric or hydraulic drive and can be powered directly or indirectly. Electric-drive machines generate less heat and are more efficient (Fowler, 2011); hydraulic-drive shredders are more portable and may be less affected by high-load conditions or unshreddable material.

Rotary shredders substitute high torque and low speed (5–50 rpm) for the low torque and high speed of hammer mills. As a result, the power use of double-rotor shredders such as that shown in Figure 5.7 is similar to that of hammer mills. However, the lower speed produces a more uniform product, reduces the generation of dust and fines, and makes shredding much less noisy. The shreds generated by these devices tend to be less wadded up than those coming from a hammer mill; this makes decoating and paint removal more effective. Rotary shredders also generate less dust and vapor, reducing the risk of explosion (Graveman, 2009). One approach to reducing the high power cost is the use of a single-rotor shredder, in which the metal is held against a stationary plate while being shredded. However, the need for higher capacities has encouraged the use of two- and even four-rotor shredders (Fowler, 2011). This is particularly the case if the rotary shredder is being used to process white goods (see Chapter 3). Table 5.1, taken from the assessment by Evans and Guest (2000), compares the use of different shredding devices for UBCs.

TABLE 5.1
Comparison of Shredder Technologies

	Hammer Mill	Impact Crusher	Contrashear	Monoshear
Shredding action	Impact	Impact	Shear	Shear
Compaction of shredded product	Medium	Medium	High	Low
Size uniformity of shredded product	Poor	Poor	Poor	Good
Generation of fines	High	High	Low	Low
kWh/1000 kg for shredding loose UBC bales to 5 mm pieces	42.0	N/A	88.0	27.0
Cost of electricity (USD/1000 kg @ $0.06/kWh)	$2.78	N/A	$5.82	$1.79
Cost of replacement cutters (USD/1000 kg)	$3.32	N/A	$1.61	$2.82
Total running cost (USD/1000 kg)	$6.10	N/A	$7.43	$4.61
Noise	Poor	Poor	Good	Good

Source: Evans, R. and Guest, G., *The Aluminium Decoating Handbook,* Stein Atkinson Stordy Ltd., Wolverhampton, U.K., June, 2000, http://infohouse.p2ric.org/ref/26/25275.pdf. With permission.

SORTING (NIJKERK AND DALMIJN, 1998)

Hand sorting: As the name suggests, this is the process of going through a feed stream and removing desirable items by hand, putting them into separate bins for later processing. It was the first method ever used for sorting garbage or scrap and continues to be popular.

The most widespread use of hand sorting is the initial processing of household waste. The collection of aluminum scrap from this source was described in Chapter 3. The blue bag used to separate recyclables from other garbage is a form of hand sorting. Many cities with MRFs perform their own hand sorting once garbage reaches the main processing facility (Kessler Consulting, Inc., 2009). The material removed from the garbage varies among facilities but almost always includes metal scrap, in particular copper, brass, and aluminum.

Another common location for hand sorting is the product streams from shredders (Gesing, 2006; Mutz et al., 2003; Wens et al., 2010). The magnetic fraction can contain items such as motors and other items with attached contaminants. Removing these items helps reduce impurity levels in the ferrous scrap. The nonmagnetic fraction is also frequently hand sorted, even if other sorting devices are used. This helps remove items of similar density but different composition. An example of this is the pieces of copper wire in aluminum scrap. The plastic or rubber coating of these wires gives them a density similar to that of aluminum, and the conductivity of the copper makes it responsive to eddy current separation. Hand sorting to remove the wire accomplishes a separation not readily achievable by machine. Separating aluminum from copper wire in cable scrap is also typically a hand sorting process (Taylor, 2010).

Hand sorting is obviously labor intensive, in addition to being unpleasant and possibly unhealthy work. However, in parts of the world where labor costs are low, it is still a preferred method for scrap sorting. In fact, much of the nonferrous scrap generated by automotive shredding in North America and Europe is sent to Asia (Gesing et al., 2010; Minter, 2006), where hand sorting is less expensive than the automatic processes used in more developed nations. It is still a preferred method for garbage sorting as well (Khullar, 2009; Minter, 2010; Wilson et al., 2009).

Hand sorters use several visual recognition tools to distinguish desirable items in a feed stream (Gesing et al., 2006; Mutz et al., 2003). Color is the most important, allowing separation of metallic items from nonmetallic and allowing the removal of copper, brass, and lead from other metals. Skilled hand sorters can distinguish cast from wrought aluminum and aluminum from magnesium. Shape is also important in the sorting of garbage, since tableware is usually made of stainless steel, cans are made of aluminum or steel, plastic milk cartons are produced from high-density polyethylene, etc. A significant challenge in aluminum recycling is the recovery of aluminum from composite aseptic packaging such as juice boxes. The juice boxes consist mostly of paper and plastic and will not be separated from other garbage by methods that depend on density difference or metallic properties. Visual recognition and removal can create a supply of this type of packaging large enough for a separate treatment process.

Hand sorting has limitations. Metal parts that are painted or plated with another metal can be misidentified. Some parts are produced using a choice of alloys, so shape recognition may not be sufficient to segregate by type of metal. Finally, if labor costs are too high (Minter, 2011; Wens et al., 2010), the value of recovered materials may not justify the cost of sorting (particularly in MRF operations). As a result, efforts are being made to develop sorting tools that can replace hand sorting (Gesing et al., 2010). These will be described later.

Air classification: Metal scrap often contains high levels of low-density impurities. Municipal waste consists mostly of waste paper and plastic; the shredder fluff from auto shredding contains small amounts of metal and large amounts of organic material from the carpets, dashboard, etc. A more recent source of scrap metal is electronic scrap (Toto, 2012), in particular the circuit boards. These consist mostly of plastic, which must be separated from the metal.

When the fraction of low-density material in the raw material is much higher than that of the metal to be recovered, air classification devices become useful. These devices use an upward flow of air to lift the paper and plastic away from the metal, separating the two. A large number of air classifiers have been developed; the reader is encouraged to see Nijkerk and Dalmijn (1998) for a complete listing. Two of the more important will be described here.

Figure 5.9 illustrates a *zigzag* or *Z-box* vertical air classifier, widely used for processing municipal solid waste. Air blown in through the bottom pushes low-density material (the Light Fraction) up and out, while allowing higher-density metal scrap (the Heavy Fraction) to fall to the bottom. The zigzag design improves efficiency

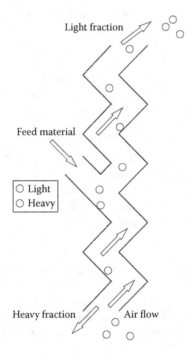

FIGURE 5.9 A zigzag air classifier. (From Nijkerk, A.A. and Dalmijn, W.L., *Handbook of Recycling Techniques,* Nijkerk Consultancy, The Hague, the Netherlands, 1998. With permission.)

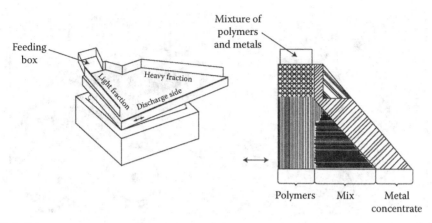

FIGURE 5.10   An air table. (From Nijkerk, A.A. and Dalmijn, W.L., *Handbook of Recycling Techniques,* Nijkerk Consultancy, The Hague, the Netherlands, 1998. With permission.)

by agitating the contents. This keeps paper and metal pieces from sticking together or blocking each other's path. Zigzag classifiers have also been applied for metal recovery from auto shredder fluff (Phillips, 1996).

Figure 5.10 depicts an *air table,* widely used for metal recovery from shredded wire and cable (Phillips, 1996; Sullivan, 1985). This material has a small particle size and thus needs to be processed more gently than the large pieces fed to vertical air classifiers. Unseparated material fed in the box at the top left is gently shaken as it moves along the sloping table toward the discharge at the far end. At the same time, air blown through small holes in the table lifts the lighter plastic and rubber in the feed, moving it toward the near side of the table. Metal scrap is less affected by the air and continues toward the far end. Collecting separate fractions from opposite sides of the discharge end produces two *concentrates.* A third *middlings* fraction, with too much plastic and metal to use as either fraction, is recycled to the feed box for another try. Shredded electronic scrap is also separated using air tables (Cui, 2005), as are fines from auto shredding operations (Eriez, 2012).

Magnetic separation: This is one of the simplest steps in the upgrading of aluminum scrap. Its purpose is to remove the iron and steel and nickel-based alloys if any are present. A variety of separators are available for performing this task (Manouchehri, 2007); the most common are *drum magnets* and *overhead belt (suspension) magnets* (Figure 5.11). A drum magnet features a stationary magnet around which a drum rotates. Nonmagnetic material falls off the drum and is collected on a conveyor running below; magnetic material sticks to the drum and continues to do so until the drum rotates past the field generated by the magnet. Then it too falls off and is removed by a separate conveyor. In a belt magnet, the drum is replaced by a conveyor belt running over the stationary magnet (Steinert, 2003b). Magnetic particles are pulled off a conveyor running below this assembly by the magnetic field and stick to the upper belt. When this belt moves away from the magnetic field, the magnetic particles drop off and are collected on a separate conveyor. Nonmagnetic particles are unaffected by the magnet and remain on their original conveyor belt.

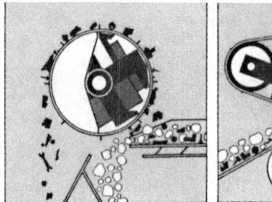

FIGURE 5.11 Drum and overhead-belt magnetic separators. (From Nijkerk, A.A. and Dalmijn, W.L., *Handbook of Recycling Techniques,* Nijkerk Consultancy, The Hague, the Netherlands, 1998. With permission.)

Because the belts are subject to wear, drum magnets are favored when the relative amount of ferrous material to remove is heavy (Steinert, 2003a).

Figure 5.12 shows a typical strategy for magnetic separation of auto shred (Milton, 2010; Ness, 1984). In this case, the magnetic fraction from a drum separator drops onto a conveyor belt that carries it under a second drum separator (Otaki, 2009). The purpose of this is to *clean* the magnetic material of any paper or other nonmagnetic material that might have inadvertently stuck to it when it was pulled onto the first drum. In plants trying to remove the ferrous impurity from the nonferrous fraction (such as UBC recyclers), it is the nonferrous fraction instead that is passed under

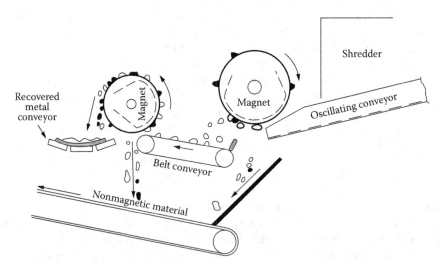

FIGURE 5.12 Magnetic separation of auto shred. (From Ness, H., The automobile scrap processing industry, in *Impact of Materials Substitution on the Recyclability of Automobiles,* ASME, New York, p. 39, 1984. With permission.)

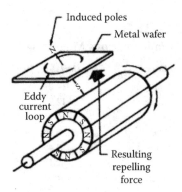

FIGURE 5.13 Formation of eddy currents by alternating magnetic fields. (From Walker National, Inc., (2011), http://www.walkermagnet.com/Collateral/Documents/English-US/Eddy Current Separation Equipment.pdf. With permission.)

a second magnet for cleaning. Iron contamination of scrap aluminum is increasingly unacceptable, so aluminum scrap may wind up being passed under magnetic separators both during upgrading and before charging to a furnace for remelting.

Eddy current sorting: Figure 5.13 illustrates a conductive particle (such as a piece of scrap) entering the magnetic field generated by a permanent magnet. The magnetic field causes an electric current to flow in the conductive particle (Milton, 2010). This current is known as an *eddy current*. The magnitude of the current is determined by the strength of the magnetic field, the electrical conductivity of the particle, and by the size and shape of the particle. The eddy current in turn generates a secondary magnetic field around the particle, aligned with the primary magnetic field generated by the permanent magnet.

When this particle passes into the magnetic field generated by a magnet of reverse polarity, the secondary magnetic field is now opposed to the primary magnetic field. This creates a repulsive *Lorentz force* that acts to flip the particle and realign it with the new primary magnetic field (Meier-Staude et al., 2002). As the particle flips, it deflects from its original direction. The degree to which it deflects is determined by several factors, but the most significant is the material itself. Table 5.2 lists several materials according to their value of σ/ρ, where σ is the electrical conductivity and ρ is the density. Aluminum and magnesium have high conductivity and low density. This means that they can be easily deflected from a given path by passing them through a series of magnets of alternating polarity. Copper is also very conductive, but its density is much higher. As a result, a piece of copper scrap is less affected by alternating fields. Stainless steel is less affected still. This is important, since stainless steel is not easily distinguished from aluminum by hand sorters and is also not removed by magnetic separators. Finally, plastic and paper are nonconductors and so are unaffected by magnetic fields. Changing magnet polarities will have no effect on them at all.

Figure 5.14 illustrates the *standard* approach for eddy current separation (ECS). Feed is placed on a conveyor, which passes over a rotating drum containing permanent magnets arranged in alternating polarity. As the particles on the

TABLE 5.2

σ/ρ for Various Materials

Material	σ/ρ (M²/Ω-kg 10³)
Aluminum	14.0
Magnesium	12.9
Copper	6.7
Silver	6.0
Zinc	2.4
Gold	2.1
Brass	1.8
Tin	1.2
Lead	0.45
Stainless steel	0.18
Glass	0
Plastics	0

conveyor belt pass over the magnets, the changing polarity causes them to deflect from their original path. The deflection causes conductive particles to fly off the end of the conveyor; the higher the value of σ/ρ, the farther the particle goes. Nonconducting particles, or those with a low value of σ/ρ, simply fall off the end of the conveyor. Ferrous particles stick to the magnet as before, but as the conveyor moves them away from the drum, these particles fall off, eliminating the need to clean them off the belt.

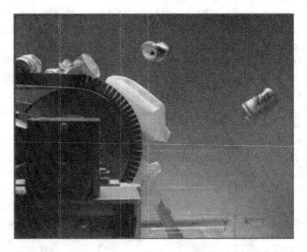

FIGURE 5.14 Conveyor-type eddy current separator. (From Milton, D., Advances in downstream physical separation, presented *at ISRI Operations Forum*, Phoenix, AZ, October 4, 2010, http://www.recyclingtoday.com/FileUploads/file/recycling/1101_RT_DownstreamSeparation_Presentation.pdf. With permission.)

In addition to the material, other factors influence how far particles fly off the end of a conveyor-type ECS (Ruan and Xu, 2012):

- Field strength of the magnets. Larger magnets improve the separation efficiency of an eddy current unit but add to the cost.
- The rotation speed of the drum (increases the force) and the speed at which the material is fed to the drum (decreases).
- The radius of the drum. Other variables being equal, a larger drum reduces throwing force.
- Mass of the particles. Larger particles generate larger eddy currents, which generate larger secondary magnetic fields, which results in larger Lorentz forces. As a result, heavier pieces of aluminum will fly farther than smaller ones. ECS works best when handling uniform-sized feed.
- Particle shape. The more nonspherical the particle, the more eddy current is generated. As a result, a piece of plate will fly farther than a ball bearing of equivalent mass.
- Particle bulk density. Hollow pieces have a larger equivalent *size* than bulkier particles. As a result, using an ECS to separate UBCs will require weaker magnets than needed for separating aluminum castings.

The conveyor-type separator shown in Figure 5.15 is generally known as a *concentric* ECS (Milton, 2010). It is the most widely used type of ECS, but it has one significant disadvantage. Ferrous pieces that happen to drop to the inside of the belt will stick to the drum as it rotates. Squeezed between the belt and drum, these pieces will eventually penetrate the drum and wedge between the drum and the rotating magnets, causing considerable damage. Figure 5.15 compares the concentric design with the more recent *eccentric* ECS (Steinert, 2012). In units of this type, the rotating magnets are mounted nonconcentrically to the conveyor pulley. This leaves a gap at the bottom between the pulley head and the support drum for the magnets. Ferrous

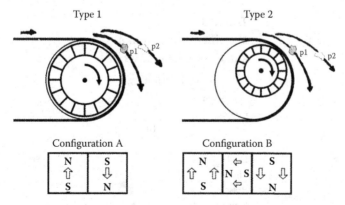

FIGURE 5.15   Conventional and HVSC magnet arrangements for ECS. (From Nijkerk, A.A. and Dalmijn, W.L., *Handbook of Recycling Techniques,* Nijkerk Consultancy, The Hague, the Netherlands, 1998. With permission.)

pieces landing on the inside of the belt are pulled away by the conveyor, improving the likelihood that they will eventually fall off before causing any damage. Whether concentric or eccentric ECS is more effective is a matter of debate.

Figure 5.15 also illustrates an alternative way of arranging the magnets in an ECS. Most units use the arrangement shown at the bottom left, in which magnets of reverse polarity are alternated. However, a more recent design (bottom right) by Huron Valley Steel Corp. (HVSC) rotates the polarity by 90° in each adjacent magnet, rather than 180°. This produces a stronger magnetic field at the surface and a sinusoidal field pattern that improves the performance of the separator.

The biggest limitation of ECS has been particle size. For the most part, scrap pieces smaller than 10 mm cannot be sorted using conventional devices. This makes their use difficult in treating shredded electronic scrap. Previous options for sorting of fines included hand sorting and air tables, but the small amount of metal recovered made these processes difficult to justify (Manouchehri, 2007). As a result, fines have often wound up being landfilled along with nonmetallics in the scrap. Advances in ECS over the past 10 years include techniques for better recovery of fines. High rotation frequencies help achieve this result, and eccentric rotor placement may help (Fischer, 2007; Steinert, 2012).

Heavy-media separation (HMS). Table 5.3 lists the densities of aluminum and common contaminants in aluminum scrap. A simple way of separating these materials from the aluminum is to put the mixed scrap in a fluid with a density between that of aluminum and the contaminant. Materials with a density less than that of the fluid will float, while those denser than the fluid will sink. HMS uses *designer fluids* to allow such separations. There are three types of fluid.

The first type is water (specific gravity = 1.0). All metals and some plastics have densities higher than 1.0, but wood, paper, shredder fluff, and most plastics are lighter than water and will float. Municipal recovery facilities often use water-based separation devices to separate a metal-rich fraction from other garbage. Figure 5.16 shows a *rising current separator* or *water classifier*, in which water pumped up through a column floats a low-density fraction, which overflows

TABLE 5.3
Densities of Aluminum
and Contaminant Metals

Material	Density (g/cm³)
Aluminum	2.7
Copper	8.97
Silver	10.5
Zinc	7.14
Lead	11.3
Tin	7.28
Brass	8.4
Stainless steel	7.9
Magnesium	1.87

FIGURE 5.16 A rising current hydraulic separator. (From LPT Group, Inc., http://www. lptgroupinc.com/pdf/lpt_litesoutclassifier.pdf. With permission.)

at the top (Phillips, 1996), and recovers a high-density fraction from the bottom. Water-only HMS is rare for scrap with a high metal content.

The addition of a dissolved salt can create fluids with specific gravities higher than 1.0 (Phillips, 1996). Calcium chloride, calcium nitrate, and potassium carbonate are frequently used for this purpose. The resulting solution has a specific gravity higher than that of polyvinyl chloride (PVC), the plastic most commonly used to coat wires. As a result, dissolved-salt solutions are most commonly used in wire-chopping operations. Dissolved-salt solutions with a specific gravity higher than that of magnesium are difficult to prepare. Because of this, these solutions are limited to separating nonmetallics from metals in the feed material.

The most common type of fluid is suspended-solid slurry (Manouchehri, 2007), in which small particles of a dense solid are suspended in water. The solid is usually either ferrosilicon (84% Fe, 16% Si) or magnetite (Fe_3O_4). These materials are ferromagnetic, which allows them to be easily removed from the solution. They are also inexpensive, corrosion resistant, and nonhazardous to the environment if lost or disposed of. By adding different levels of ferrosilicon or magnetite, the specific

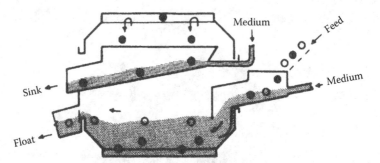

FIGURE 5.17 A drum-type heavy-media separator. (From Sepor, Inc., http://www.sepor.com/new/HeavyMed.pdf. With permission.)

gravity of the suspension can be varied to values greater than 3.0. This allows the industrial use of two heavy-media suspensions. The first, with a specific gravity of about 2.0, separates aluminum and heavier materials from the magnesium, plastic, and other light materials in the feed stream. The second, with a specific gravity higher than 2.4, recovers aluminum in the floats while other metals remain in the sink fraction.

Figure 5.17 shows a typical drum separator used for HMS (Manouchehri, 2007). HMS is widely used for recovering aluminum from mixed scrap, but it has disadvantages:

- Sludge and dirt in the feed can affect the specific gravity of the suspension, making it difficult to control.
- Composite materials with intermediate specific gravities may report to the wrong fraction. An example is plastic-coated copper wire, which has an overall specific gravity low enough to be confused with aluminum. Hollow particles also have an incorrect apparent density and wind up in the wrong product stream.
- Smaller particles settle poorly in heavy media, so HMS is generally limited to pieces larger than 16 mm.
- As a wet process, HMS requires drying the products following separation. It also requires washing the media off the product and recovering it, an additional capital expense.
- The loss of suspension media with the product (typically 4–5 kg/tonne of processed material) may be a source of contamination. This is a particular problem if minimizing the iron content in aluminum scrap is important. Media lost in the product also has to be replaced, which increases operating costs.

Because of this, efforts have been made to develop a dry heavy-media system to replace the suspensions (Phillips, 1996; SGM Magnetics, 2009). The dry system (Figure 5.18) uses a fluidized bed of sand kept in motion by electric vibrators. By varying the type of sand used, the specific gravity of the bed can be varied between 2.3 and 4.6. Scrap fed to the fluidized bed sinks or floats according to its specific gravity. Dry HMS is not useful for separating aluminum from lighter materials

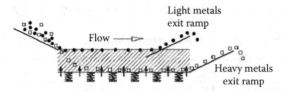

FIGURE 5.18 Dry (sand-based) heavy-media separation system. (From SGM Magnetics, SGM Magnetics, *Aluminum separator type Sandjet Data Sheet,* SGM Magnetics, Sarasota, FL, 2009. http://www.holgerhartmann.fi/sites/default/files/sandjet_english_09.pdf. With permission.)

but can be used to separate it from heavier metals. Whether this approach is cost-effective is still uncertain. HMS is currently losing favor to sensor-based sorting methods, except for wet shredding flowsheets (Gesing and Harbeck, 2009).

SENSOR-BASED SORTING

The sorting techniques described in the previous section are all familiar elements of flowsheets for recovering aluminum and other metals from comminuted scrap. However, as the previous discussion has pointed out, all have limitations. These limitations have become more apparent with the need for improved results from sorting. Gesing et al. (2010) list eight types of separation desired from a sorting scheme:

- *Separating ferrous material from nonferrous and nonmetallic materials*: This is the simplest separation to accomplish, but the recent need to produce a purer ferrous scrap stream has made it more difficult. Magnets will not recover many grades of stainless steel, and these are a significant potential contaminant in scrap aluminum.
- *Separating nonferrous metal from nonmetallic materials*: The adoption of ECS has made this easier, but separation of non-liberated metal pieces from those without attachments is a challenge.
- *Separation of metallic from nonmetallic fines*: New ECS units have made this more feasible, but finding economically viable ways of doing this is still difficult. Even hand sorting of fines is difficult to economically justify.
- *Removing contaminants from scrap aluminum*: The biggest concern is stainless steel, but plastic from e-waste and small copper wires also find their way into aluminum scrap.
- *Separating light from dense nonferrous metal*: Several techniques (ECS, HMS, hand sorting) accomplish this, but each has limitations that recommend replacement.
- *Separating aluminum from magnesium*: These two light metals respond the same way to several sorting techniques, and the growing levels of magnesium in automotive and electronic scrap make it increasingly difficult to ignore. Dissolved-salt HMS is effective for separating the two, but most scrap processors would prefer to avoid it. Hand sorting is somewhat effective, but the need to produce *cleaner* wrought aluminum scrap compatible

with production of low-Mg alloy types is encouragement for the development of something better.

- *Separating liberated aluminum from aluminum scrap pieces with attachments and coatings*: It makes sense to decoat only that scrap with substantial amounts of organic material in it. Sweat melting should be limited to
  scrap actually attached to pieces of iron.
- *Separating aluminum alloy families from each other*: The problem of aluminum alloy tolerance was described in Chapter 2. The use of old scrap to
  produce wrought alloy would be more feasible if aluminum scrap could be
  sorted by alloy series: 2000-series scrap in one pile, 4000-series scrap in
  another, and so on. As the popularity of wrought aluminum continues to
  grow relative to that of cast aluminum (see Chapter 3), this type of separation is becoming increasingly desirable.

Over the past 15 years, several *sensor-based* technologies have been developed that
promise to be more effective at performing these types of separation than the previous generation of methods. Figure 5.19 illustrates the first of these, *induction* or
eddy current coil sensors (Milton, 2010; Steinert, 2003c). The sensor, placed underneath a conveyor belt carrying nonmagnetic shredded scrap (Figure 5.20), features
an oscillator that emits a high-frequency electromagnetic field. This field generates
an eddy current in conductive material passing over it. The eddy current in turn
causes a loss of energy in the oscillator, which is detected by the sensor. Detection
causes the activation of a circuit that leads to ejection of the conductive piece from
the conveyor using either an air jet or mechanical fingers. The primary purpose of
induction sorters is to recover metal from the nonmetallic stream created by ECS.
However, these sensors can also detect the difference between paramagnetic stainless steel and diamagnetic metals such as aluminum and magnesium (Gesing, 2006).
The separation of stainless steel from nonferrous scrap is difficult to accomplish, and
this is a valuable attribute. If the particles on the conveyor belt are of similar size,
high-sensitivity induction sorting can also be used to separate pieces of wire from
nonmetallic material such as shredder fluff.

Hand sorters have long used differences in the color of scrap pieces to separate copper and brass from other metals. High-resolution charge-coupled device
technology is now used to perform this function automatically, again linked to air-
jet or mechanical devices for ejecting particles from a conveyor into the right pile

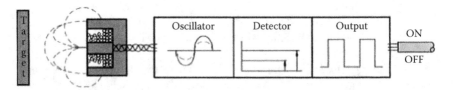

FIGURE 5.19 Principle behind induction sorting. (From Milton, D., Advances in downstream
physical separation, presented *at ISRI Operations Forum*, Phoenix, AZ, October 4, 2010, http://
www.recyclingtoday.com/FileUploads/file/recycling/1101_RT_DownstreamSeparation_
Presentation.pdf. With permission.)

FIGURE 5.20 Induction sorting sensors underneath a conveyor. (From Milton, D., *Advances in downstream physical separation, presented at ISRI Operations Forum*, Phoenix, AZ, October 4, 2010, http://www.recyclingtoday.com/FileUploads/file/recycling/1101_RT_DownstreamSeparation_Presentation.pdf. With permission.)

(Ridall, 2009). While color separation cannot separate aluminum from other *white* metals such as magnesium, it can potentially be used to separate painted from bare scrap (Gesing, 2006). This reduces the amount of unnecessarily decoated material and thus the cost.

The use of x-ray technology to identify scrap pieces has become a significant element in sorting flowsheets over the past decade. X-ray transmission (XRT) through scrap pieces is blocked in proportion to the thickness of the piece and the atomic number of the metal and can be used to distinguish aluminum from heavier metals such as copper and stainless steel without being affected by irregular shapes the way that HMS can (Gesing, 2006; Steinert, 2007). As a result, XRT is replacing HMS in some sorting processes (Fischer, 2007). Dual-energy XRT (DEXRT) eliminates the effect of particle thickness and makes possible the separation of aluminum from magnesium, another difficult challenge. It might also be possible to use XRT to separate wrought aluminum scrap from cast, given the high concentration in the latter of heavier alloying elements copper, manganese, silicon, and zinc.

A second effect on x-ray absorption by substances is the emission of fluorescence (XRF) in response. The intensity of the fluorescence is a function of wavelength, characteristic of the chemical element; each element has a *signature* that can be detected (Donovan, 2011). The higher the concentration of a particular element in a material, the greater the level of that element's signature that will be measured. As a result, alloys have an overall XRF signature as well, and comparing the response from XRF of scrap pieces against these signatures can be used as an identification tool. This has been used for some time for field identification of scrap metal class (stainless steel, titanium, etc.) but more recently has been applied for identification of specific alloys (Gaustad et al., 2012; Gesing, 2007). Online systems have been developed and are being integrated into scrap sorting flowsheets (Fischer, 2007). Limitations to XRF as a scrap sorting tool include relatively slow speed, equipment cost, and a relative inability to identify low-atomic-number elements such as aluminum and magnesium. The heavy elements in paint and other coatings can also interfere with XRF analysis. As a result, XRF is not commonly used to identify aluminum alloys in scrap, although it may be useful for separating alloy series.

A second spectrographic analysis approach for alloy identification may have better potential for aluminum alloy sorting. Laser-induced breakdown spectroscopy (LIBS) involves illuminating a spot on a piece of scrap with a focused laser beam (Gaustad et al., 2012; Gesing, 2006, 2007). This causes the scrap surface to fluoresce. Optical emission spectroscopy (OES) of the fluorescence photons is used to produce a quantitative analysis of the alloy composition, which is then compared with alloy specifications. Once identified by alloy, the scrap piece can then be diverted to the proper scrap bin, separating it from scrap of differing compositions. As with XRF, surfaces need to be free of paint, in order to avoid analyzing the paint instead of the alloy. They also need to be decoated, since some coatings (such as zinc phosphate) contain alloying elements. Calibration of the detector is also vital to getting accurate results. LIBS has the advantage of being able to detect and analyze light elements such as aluminum, magnesium, and silicon, making it more applicable for aluminum scrap identification and sorting. Figure 5.21 illustrates the results of LIBS analysis of six common aluminum alloys (Gesing et al., 2010). The ability to distinguish between types of alloy is apparent.

Commercial LIBS analyzers for aluminum scrap processing are designed for single-shot analysis of cleaned surfaces (Gesing et al., 2010). Particles are presented to the laser one at a time, at an average rate of 40/s. LIBS processing is performed on shredded scrap with 1–15 cm particle size, moving at 3 m/s. This allows processing of up to seven tonnes of material per hour. The first industrial facility of this type was

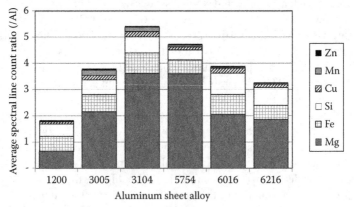

FIGURE 5.21 LIBS *signatures* of common aluminum alloys. (From Gesing, A., et al., Advanced industrial technologies for aluminium scrap sorting, presented at *Aluminium-21/ Recycling*, St. Petersburg, Russia, 2010. With permission.)

TABLE 5.4

Types of Sorting Performed on Scrap Aluminum and Available Sorting Techniques

Type of Sorting	Techniques
Steel from nonferrous material	Drum magnet, belt magnet
Nonmagnetic metal from nonmetal	Screen, air separator, eddy current separator, induction sorter
Metallic from nonmetallic fines (<9 mm)	Jig, eddy current separator
Aluminum from other materials	Magnetic head pulley, eddy current separator, hand sorting
Light from dense nonferrous metals	Heavy-media separation (wet or dry), eddy current
Aluminum from magnesium	separation, hand sorting, color sorting, XRT sensor
Aluminum with attachments from liberated aluminum	Hand sorting, XRT sensor
Aluminum alloys from other aluminum alloys	XRF sensor, LIBS sensor

recently commissioned by Huron Valley Steel Corp. in 2004 in the United States and has been in operation since.

Table 5.4, derived from the summary by Gesing et al. (2010), illustrates the processing options for the different types of sorting of aluminum scrap. It is likely that sensor-based technologies will play a more prominent role in the future, and hand sorting and HMS will continue to decline in significance.

THERMAL PROCESSING

Decoating: Aluminum and its alloys often come coated with an organic material of some sort. A variety of lacquers and polymers are used to coat extruded material, most notably can stock and foil. 2xxx and 7xxx alloys are often painted, and turnings and borings from machining operations bring with them substantial levels of grease and oil. Table 5.5 lists the coatings of different aluminum products (Evans and Guest, 2000) and the weight fraction of the product consisting of organic material.

Typical practice until the 1980s was to charge coated or oily material directly into the melting furnace along with the rest of the scrap. This had several harmful effects (McAvoy et al., 1990):

- The organic material would burn upon being charged to the furnace. The resulting smoke and soot was a workplace health hazard and difficult to remove from the plant off-gas.
- The burning coating heated the metal underneath, causing it to oxidize as well (Rossel et al., 1997). Figure 5.22 shows the melt loss from various types of coated and uncoated scrap charged directly to a furnace (McAvoy et al., 1990). It can be seen that painted or lacquered scrap oxidizes more than bare scrap regardless of alloy. Melting under a salt flux helps some, but melt losses from laminated material are still over 5%.

TABLE 5.5

Scrap Types and Their Coatings

Scrap Type	Coating/Structure	Coating (wt.%)
Aerosols	Lacquers, paints	2–3
Capstock	Polymers, lacquers	30
Converted foil	Inks, lacquers	7
DSD	Various	<85
Epoxy strip	Epoxy, paint	6
Frag	Oil, paint, polymers	<4
Lid stock	Vinyls	8
New lithographic plate	Phenolic resin, paper	<4
Mill foil	Rolling oils	<10
Paper laminate	Paper, inks	50–60
Plastic laminate	Polymers, inks, wax	40–90
Painted sidings	Paints	3
Tube laminate	Polymers, lacquers	70
Turnings	Cutting oils	<20
UBC/new can stock	Lacquers, paints	2–3
Used lithographic plate	Phenolic resin, inks	2
Window frames	Thermolacquers, resins	21

Source: Evans, R. and Guest, G., *The Aluminium Decoating Handbook,*
Stein Atkinson Stordy Ltd., Wolverhampton, U.K., June 2000.
http://infohouse.p2ric.org/ref/26/25275.pdf. With permission.

- In addition to soot and smoke, other troubling emissions from the burning
  coating included dioxins and furans. Dioxins and furans were more likely
  to form as the temperature rose and are considered especially dangerous in
  even minute quantities (Sweetman et al., 2004).

Because of this, coated or oily scrap became unwanted material in some secondary
smelters and foundries. UBCs were especially unpopular (van Linden, 1994), since
they were difficult to submerge in the melt and even more likely to be a high melt-
loss material. However, the growing popularity of aluminum beverage cans, and
the development of specialized facilities to recycle them, encouraged research into
finding better ways of removing oil and lacquer from scrap surfaces.

A decoating process has several goals:

- Complete removal of the organic material from the scrap surface
- Minimized oxidation of the scrap surface
- Low emissions of harmful by-products, such as volatile organic compounds
  (VOCs), dioxins/furans, and nitrous oxides (NO_x)
- Use of the fuel value of the organic material to make the process *autother-*
  mal, capable of supplying its own energy needs
- Minimal capital and operating cost

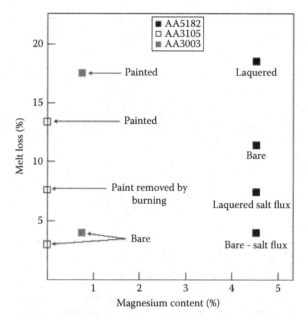

FIGURE 5.22 Effect of alloy type and coating on melt loss in fluxed and flux-free remelting. (From McAvoy, B. et al., The Alcan decoater process for UBC decoating, in *2nd International Symposium on Recycling of Metals Engineered Materials*, van Linden, J.H.L., Stewart, D.L. Jr., and Sahai, Y., Eds., TMS-AIME, Warrendale, PA, p. 203, 1990. With permission.)

Figure 5.23 (Kvithyld et al., 2002) illustrates the results of thermogravimetric analysis (TGA) of lacquered aluminum scrap as it is heated in a low-reactivity environment (Ar + 1% O$_2$). The bottom curve shows the weight decrease of the sample as the lacquer is decomposed to form gaseous species; the top curve shows three peaks representing the thermal effects of chemical reactions. The first of these peaks is thought to represent the breaking of the polymer chains from which the coating is formed. The second peak is the formation of VOCs. Mass spectrometric analysis of the off-gas from inert-atmosphere decoating of scrap reveals a large number of organic compounds; the most significant are benzene, toluene, phenol, styrene, benzaldehyde, dioctylphthalate, and bisphenol-A (Evans and Guest, 2000). The formation of VOCs leaves behind a residue on the surface known as *char*. The third peak of the top curve in Figure 5.23 represents the oxidation of this char. When scrap is decoated in an oxygen-free environment, this third peak does not appear.

The rate at which these reactions happen varies with temperature and atmosphere. Table 5.6, adopted from the results of Kvithyld et al. (2002), predicts the lifetime of coatings at different temperatures under Ar and Ar + 1% O$_2$ atmospheres. Higher temperatures have an obvious impact, particularly on the second step of the process. However, decoating at higher temperatures increases energy use, making it more difficult to operate the process autothermally. It also increases surface oxidation of the metal during the process of burning off the char. This results in higher melt

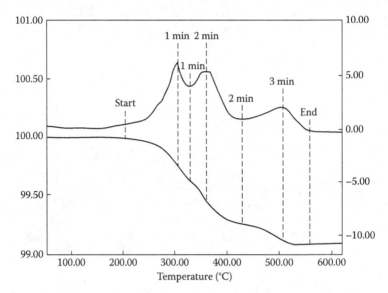

FIGURE 5.23 Thermogravimetric analysis of the thermal decomposition of the lacquer coating on UBCs. (From Kvithyld, A. et al., Decoating of aluminium scrap in various atmospheres, in *Light Metals 2002*, Schneider, W., Ed., TMS-AIME, Warrendale, PA, p. 1005, 2002. With permission.)

loss. As a result, current decoating strategy relies on prolonged exposure at a lower temperature (480°C–520°C), which minimizes metal oxidation. However, this can also result in incomplete decoating, leaving behind unoxidized char (Stevens and Tremblay, 1997).

For scrap with low organic content, decoating in the preheating chamber of a twin-chamber melting furnace is effective and eliminates the need for a separate unit (Evans and Guest, 2000). This will be further described

TABLE 5.6

Lifetime of Lacquer Coating during Thermal Decoating under Varying Conditions

	Atmosphere		
Decoating Temperature (°C)	Ar (s)	Ar + 1% O_2 (s)	Air (s)
400	21	13	7
500	0.4	0.2	0.2
600	0.02	0.01	0.01

Source: Kvithyld, A. et al., Decoating of aluminium scrap in various atmospheres, in *Light Metals 2002*, Schneider, W., Ed., TMS-AIME, Warrendale, PA, p. 1055, 2002. With permission.

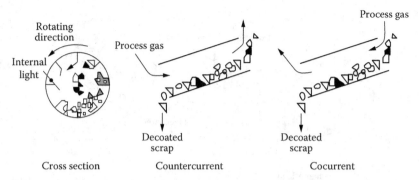

FIGURE 5.24 Rotary-kiln decoating furnace. (From Stevens, W. and Tremblay, F., Fundamentals of UBC decoating/delacquering for efficient melting, in *Light Metals 1997*, Hughes, R., Ed., TMS-AIME, Warrendale, PA, p. 709, 1997. With permission.)

in Chapter 8. However, high-organic scrap requires more time than a preheating chamber can afford. To effectively decoat this material, a separate furnace is needed. Figure 5.24 depicts the predominant type of decoating furnace, a rotary kiln. The kiln utilizes the second of the basic strategies for heating scrap, mixing the incoming scrap with hot *process gas*. This gas consists of VOCs from the delacquering process, burned with excess air to produce a gas generally containing 6%–8% O_2. (Oxygen levels above 10% encourage excess oxidation; levels under 4% make it difficult to oxidize the char.) The kiln is equipped with internal flights that help *throw* the scrap through the gas, improving heat transfer and helping the scrap heat faster.

Three types of rotary kiln are used. In the first, process gas is added at the same location as the incoming scrap and flows *cocurrent* with the charge (Evans and Guest, 2000). The large temperature difference between the two incoming streams ensures higher heat-transfer rates and quicker heating of the scrap. This in turn reduces the amount of char generated. However, much of the oxygen in the process gas reacts with the VOCs during heating, so less is available to react with the char. In a *counter-current* rotary kiln, the process gas and scrap are added at opposite ends and flow opposite to each other. This allows more heat to be removed from the gas, improving thermal efficiency. It also means more oxygen is available in the gas to react with the char. However, the smaller temperature differences between gas and scrap mean that heat-transfer rates are lower, and the scrap heats at a slower rate. As a result, there is more char to be oxidized than cocurrent rotary kilns generate. The IDEX furnace attempts to achieve the advantages of both types by introducing the gas through a central coaxial tube in a cocurrent direction. The temperature difference between the gas and scrap is sufficient to encourage heat transfer by radiation. When the gas exits the tube at the discharge end of the kiln, it retains sufficient oxygen content to burn off the char. It then flows back through the scrap, transferring more heat to the charge and improving thermal efficiency. The extra complexity increases capital costs, however.

Some scrap is still decoated using a *packed-bed reactor* or *belt decoater*. Figure 5.25 shows the basic design of this furnace, which heats coated scrap to a

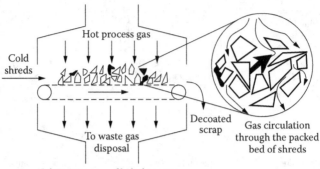

Schematic view of belt decoater

FIGURE 5.25 Belt decoating furnace. (From Stevens, W. and Tremblay, F., Fundamentals of UBC decoating/delacquering for efficient melting, in *Light Metals 1997,* Hughes, R., Ed., TMS-AIME, Warrendale, PA, p. 709, 1997. With permission.)

lower temperature for an extended period. Scrap is packed in a bed on a moving wire-mesh belt (McAvoy et al., 1990). As it moves through the furnace, hot process gas is blown or drawn through the bed. As with the rotary kiln, the gas consists of combusted VOCs generated by the decoating scrap, burned with excess air to generate a gas with 6%–8% O_2. The scrap is heated to lower temperatures than the rotary kiln (480°C–520°C) and stays in the furnace for a longer time (5 min).

Proper operation of a belt decoater is tricky (McAvoy et al., 1990), because several variables must be controlled properly to get optimal results. These include the gas flow rate through the bed, the inlet gas temperature, belt speed, bed thickness, and particle size of the shredded scrap. Typical practice is for a bed thickness of 75 mm and inlet gas temperatures of 500°C–600°C. The use of a traveling bed guarantees equal residence time for all the feed particles and contact with the gas for all the scrap. The thin bed depth reduces temperature gradients, encouraging consistent results.

Also available (but rarely used) is the *fluidized bed decoater*, developed by Alcan in the early 1990s (McAvoy et al., 1990). Figure 5.26 is a schematic of this type of furnace, which like the belt decoater uses longer exposures at lower temperatures for decoating. The key to this technology is the bed of inert particles that is fluidized by heated process gas fed through the bottom of the unit. The hot particles transfer heat to the solid pieces of scrap more efficiently than the hot gas alone, resulting in shorter required residence times than needed in a belt decoater. The operating temperature and rotation speed of the rotating decoater drum through which the scrap is fed allows decoating of a variety of feed types.

In operation, scrap is fed through the airlocks at the left and enters the drum, which has been perforated to allow fluidizing medium in and out. The fluidizing gas fed through the bottom is a mixture of compressed air and natural gas. The combustion of the natural gas and the VOCs generated by the decomposing coating generates heat to operate the process. In the first portion of the drum, the gas mixture is kept oxygen-poor to encourage volatilization of the coating without excessive oxidation of the scrap. As the scrap moves through the drum, the mixture becomes

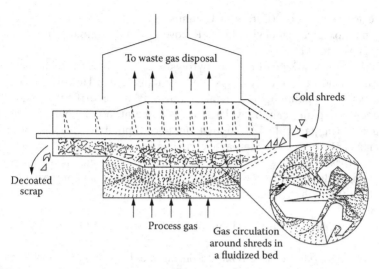

FIGURE 5.26 Fluidized-bed decoating furnace. (From Stevens, W. and Tremblay, F., Fundamentals of UBC decoating/delacquering for efficient melting, in *Light Metals 1997*, Huglen, R., Ed., TMS-AIME, Warrendale, PA, p. 709, 1997. With permission.)

progressively oxygen-rich, to encourage burning off the char. The rotation of the drum keeps the scrap from settling to the bottom. As the decoated scrap leaves the furnace, it goes through a desanding zone and then a hot media screen designed to recover the fluidizing medium. This reduces both losses of the medium and impurities in the decoated scrap.

A common element to all three types of decoater is the need to combust the VOCs in the off-gas. This provides process heat and ensures an environmentally acceptable discharge from the facility. In the fluidized bed decoater, VOCs are burnt off in a thermal decoating zone located inside the unit. In rotary kilns and belt decoaters, a separate afterburner is provided (Evans and Guest, 2000). Temperature control of the combustion is important. If the temperature is too low, the VOCs will not be completely oxidized; if the temperature level is too high, unacceptable levels of NO_x and dioxins/furans may result. In practice, afterburners often operate at around 800°C–850°C.

Pyrolysis of composite packaging: As mentioned in Chapters 2 and 3, some of the aluminum used in packaging is used in foil form, rather than the extruded can stock most commonly encountered. Aluminum foil is used in a variety of packages, most often laminated or attached to paper or plastic. The percentage of aluminum in these packages is low, and the thin aluminum foil is easily oxidized, so recycling this material with minimal melt loss is particularly difficult (Bateman et al., 1999).

When packaging of this sort is recovered, current recycling practice is to decompose the package at a paper mill through a depulping process (Steeman, 2013). The solid residue from this process is a mixture of polyethylene and aluminum, which is often incinerated or disposed of. However, a pyrolysis process has recently been developed that recovers the aluminum and decomposes the plastic to a hydrocarbon mixture for combustion (European Commission, 2011). The process operates at 450°C in an oxygen-free environment. Recovery of the aluminum from composite packaging is rare at this time.

Paint removal: Decoating and pyrolysis is effective at separating scrap aluminum from organic coatings and contaminants. However, much aluminum scrap is painted, and paint removal is more difficult. The organic component of paint can be volatilized, but paint often contains inorganic compounds (in particular TiO_2) that do not respond to thermal processing (Bateman et al., 1999). The inorganic compounds remain with the scrap and wind up in the dross when the scrap is remelted. This increases melt loss and increases the amount of dross that must be handled. Furthermore, some of the TiO_2 may be reduced by the molten aluminum:

$$3TiO_2 + 4Al\ (l) = 3Ti\ (\text{dissolved in metal}) + 2Al_2O_3\ (\text{dross}) \qquad (5.1)$$

To some degree, paint can be removed during the decoating process by abrasion, such as the tumbling action during rotary kiln operation (Bateman et al., 1999). The abrasion caused by the fluidizing medium during fluidized-bed decoating is even more effective. Research has shown that a formic acid–halogenated acetic acid solution causes the paint to swell and loosen itself from the can, making it easier to remove (Fujisawa et al., 2000). However, the practicality of this approach is doubtful, except for large quantities of heavily painted scrap.

Partial melting: In a previous section, the value of separating aluminum scrap on the basis of alloy content was discussed. One of the early attempts to partially accomplish this goal used differences between the solidus temperatures of various alloys. The solidus is the temperature where melting of the alloy begins to occur. Depending on the appropriate phase diagram, alloys with high amounts of added elements (such as casting alloys) can have a lower solidus than lightly alloyed wrought compositions (Maurice et al., 2000). This can be used to upgrade mixed-alloy scrap. As mentioned in Chapter 2, aluminum beverage cans are typically made of two wrought alloys: a low-alloy body (3004) and high-magnesium 5182 for the top and tab. The solidus temperature of 3004 is 629°C, while that of 5182 is 581°C, some 48°C lower. Using this, Alcoa designed a rotary delacquering kiln for UBCs to operate at an intermediate temperature (Bowman, 1985). The tumbling action of the kiln is sufficient to decompose the lids into small flakes, while the can bodies remain unaffected. Screening the product separates two alloys. This allows closed-loop remelting of the can bodies to produce 3004 and similar closed-loop melting of the 5182 flakes to produce separate lid stock. However, the use of partial melting to separate mixed-alloy UBC scrap ultimately proved uneconomic.

Sweat melting: While most aluminum scrap can be liberated from other metals to which it is attached by shredding, there are exceptions. Some aluminum parts are so intricately attached to iron parts that they will not be separated by ordinary shredding. Examples include transmissions, cylinder heads, and manifolds. These items are known as high-irony scrap (ISRI grade Tarry C; see Figure 4.1) because of the extremely high levels of iron (50% or more at times). Scrap of this type has too much aluminum to be classified as ferrous scrap but too much iron to be remelted by secondary smelters.

Figure 5.26 shows a traditional solution to this problem (EnviroAir Inc., 2005; Nijkerk and Dalmijn, 1998), a *sweat furnace*. Sweat furnaces can be either the rotary type shown here or a less sophisticated stationary dry-hearth design. In either case, the principle is simple. High-irony scrap fed to the furnace is heated to a temperature just above the melting point of aluminum (660°C). At this temperature, the aluminum slowly melts away (hence the term, *sweating*), leaving the higher-melting point iron behind. The liquid aluminum drains from the bottom of the furnace and is cast in the form of sows or pigs. (The product is often known as *sweated pig*; see ISRI grade Throb in Figure 4.1.) The furnaces can be operated with a variety of fuels and are also used for melting zinc away from solid aluminum in scrap radiators at lower temperatures. (This recovers aluminum as well as zinc.) Rotary furnaces do a better job of separating the melted aluminum from the iron pieces and so tend to produce a sweated pig with lower iron content (Figure 5.27).

Sweated pig is commonly sold in one of two grades, 1-1-1 or 1-1-3 (Rochester, 2009). The numbers refer to the maximum allowable percentages of magnesium, iron, and zinc in the product. There are no limits on the percentages of other alloying elements. The aluminum scrap melted in sweat furnaces is normally cast alloy, so typical copper contents might be 2.5%–3.0%, the normal silicon range is 6.0%–9.0%, and manganese might range from 0.2% to 2.0%. Sweated pig is usually sold to secondary smelters, where it is remelted and mixed with other scrap.

FIGURE 5.27   A large-scale rotary sweat furnace. (From EnviroAir, Inc., *CORECO® Rotary Sweat Furnace*, EnviroAir, Inc., Perrysburg, OH, September 8, 2005. http://www.enviroair.net/assets/Uploads/Sweat-Furnace.pdf. With permission.)

AGGLOMERATION (NIJKERK AND DALMIJN, 1998)

Balers: The aluminum scrap recovered using the technology previously described consists of loose pieces, with a very low bulk density. This makes it difficult to handle and ship and makes it more likely that it will pick up moisture during shipment. As a result, some method of densification is desirable before shipping. Figure 5.28 illustrates the operation of a baler, which compresses scrap using hydraulic rams. The baler shown here is a three-stroke unit (Zöllner, 2003). This process starts by compressing from the side perpendicular to the door; it then compresses from the top down and finally in the direction of the door. As the final compression is completed, the door opens and the finished bale is pushed out. Because baling deforms the metal during compression, metal bales are more cohesive than those made from plastic or paper and usually do not require tying with wire.

The baler shown in Figure 5.28 is a two-ram baler, which is more commonly used for processing metal scrap than single-ram units (Fowler, 2007b). Improvements in recent years include better wear resistance, improved control systems that minimize the need for an operator, and flooded suction pumps and sturdier filtration systems for better hydraulics.

While consolidation of some sort is essential prior to shipping scrap, bales are not always popular with scrap consumers. The bales are often too big to charge directly to a furnace, and bales often contain surprises—moisture, contaminants, grease, and oil. Because of this, bales are often shredded and reprocessed by the consumer upon arrival.

Briquetting: One of the more difficult types of new scrap to recycle is turnings and borings generated by machining operations (Pietsch, 1993). This type of scrap has a very large surface area/volume ratio, making it more prone to oxidation during melting. Furthermore, turnings and borings are frequently covered with grease or oil (as much as 20%–30% by weight), which is highly undesirable in remelting. In response, turnings and borings are often *briquetted*. This too is a process of compaction under pressure (30 MPa is typical), using much higher pressure and producing much smaller compacts. Briquetting reduces the effective surface area/volume ratio of the turnings, helping to reduce melt loss. It also squeezes up to 95% of the grease and oil out of the feed (Novelli, 2006), which makes subsequent processing easier (and recovers the fluids for reuse). If the scrap chips or turnings are to be shipped, briquetting can reduce its volume by up to 90%.

Figure 5.29 illustrates the two types of briquetting technology used (Pietsch, 1993). The traditional *punch-and-die* machine produces cylindrical briquettes or *pucks* with diameters of 12–15 cm and a height of roughly 10 cm. A newer approach involves feeding the material to a hydraulic *roller press*, which operates continuously and produces a strip of compacted material with a density near theoretical. Roller presses are more productive than punch-and-die briquetting machines, particularly for handling aluminum swarf; capacities range up to 2500 kg/h. However, punch-and-die machines are more cost-effective for small operations and handle more elastic turnings and borings better. The practice of shredding turnings prior to briquetting helps improve the results.

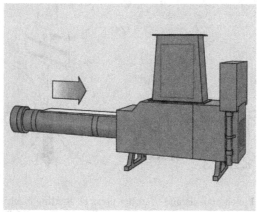

RAS I

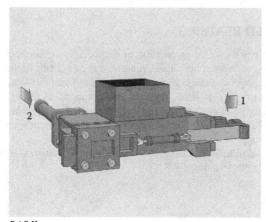

RAS II

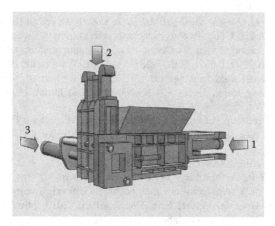

FIGURE 5.28 Operation of a three-stroke baler. (From Metso Minerals, http://www. metsominerals.com/inetMineals/MaTobox7.nsf/DocsByID/277BAEB1DECCC2FD42256B4 100451C96/$File/1229_RAS_EN.pdf. With permission.)

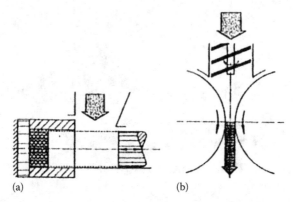

(a)   (b)

FIGURE 5.29 (a) Punch-and-die and (b) roller press briquetting machines. (From Pietsch, W., Briquetting of aluminum swarf for recycling, in *Light Metals 1993*, Das, S.K., Ed., TMS-AIME, Warrendale, PA, p. 1045, 1993. With permission.)

RECOMMENDED READING

Gesing, A., Shredder practice for preparation of nonmetallic concentrates and potential for particle sorting of these concentrates, presented at *SgS 2006*, Aachen, 29 March 2006. http://www.gesingconsultants.com/publications/Shredder-pract-sgs06-pres.pdf.

Manouchehri, H.R., Looking at shredding plant configuration and its performance for developing shredding product stream, Jernkontorets Forskning, Stockholm, Sweden, D823, March 9, 2007. http://www.jernkontoret.se/ladda_hem_och_bestall/publikationer/stalforskning/rapporter/d_823.pdf.

Nijkerk, A.A. and Dalmijn, W.L., *Handbook of Recycling Techniques*, Nijkerk Consultancy, The Hague, the Netherlands, 1998.

REFERENCES

A-Ward, *Steel Shear*, March 16, 2005. http://home.xtra.co.nz/hosts/a-ward/pdfs/SteelShearFINAL.pdf.

American Recycler, *Equipment Spotlight: Alligator Shears*, American Recycler, Perrysburg, OH, November 2003. http://www.americanrecycler.com/nov03/spotlight.html.

Bateman, W., Guest, G., and Evans, R., Decoating of aluminum products and the environment, in *Light Metals 1999*, Eckert, C.E., Ed., TMS-AIME, Warrendale, PA, 1999, p. 1099.

Bowman, K.A., Alcoa's used beverage can (UBC) alloy separation process, in *Recycling Second Recovery Metal*, Taylor, P.R., Sohn, H.Y., and Jarrett, N., Eds., TMS-AIME, Warrendale, PA, 1985, p. 429.

Cui, J., Mechanical recycling of consumer electronic scrap, Licentiate Thesis, University of Technology, Luleå, Sweden, May 2005. http://epubl.ltu.se/1402-1757/2005/36/LTU-LIC-0536-SE.pdf.

Donovan, C., X-ray vision, *Recycl. Today*, 49(7), 66, July 2011.

EnviroAir, Inc., *CORECO® Rotary Sweat Furnace*, EnviroAir, Inc., Perrysburg, OH, September 8, 2005. http://www.enviroair.net/assets/Uploads/Sweat-Furnace.pdf.

Eriez, *DensitySort Air Table*, Eriez, Erie, PA, April 20, 2012. http://rpn.baumpub.com/products/12442/densitysort-air-table.

European Commission (Environment Directorate–General), Spain: A CLEAN way to recycle drinks cartons 100%, in *Best LIFE Environment Projects 2010*, March 10, 2011. http://www.alucha.com/downloads/bestenv10.pdf.

Evans, R. and Guest, G., *The Aluminium Decoating Handbook*, Stein Atkinson Stordy Ltd., Wolverhampton, U.K., June 2000. http://infohouse.p2ric.org/ref/26/25275.pdf.

Fischer, T., Equipment focus: Downstream separation systems, *Scrap*, 64(6), 113, November/December 2007.

Fowler, J., Equipment focus: Alligator shears, *Scrap*, 60(3), 51, May/June 2003.

Fowler, J., Equipment focus: Automobile shredders, *Scrap*, 64(1), 81, January/February 2007a.

Fowler, J., Equipment focus: Balers, *Scrap*, 64(4), 91, July/August 2007b.

Fowler, J., Equipment focus: Low–speed shredders, *Scrap*, 68(5), 129, September/October 2011.

Fujisawa, K., Kogishi, T., Oosumi, K., and Nakamura, T., A basic study on development of a swell–peeling method in UBC recycling system, in *4th International Symposium on Recycling of Metals and Engineered Materials*, Stewart, D.L., Stephens, R., and Daley, J.C., Eds., TMS-AIME, Warrendale, PA, 2000, p. 835.

Gaustad, G., Olivetti, E., and Kirchain, R., Improving aluminum recycling: A survey of sorting and impurity removal technologies, *Resour. Conserv., Recycl.*, 58, 79, 2012.

Gershow Recycling, http://www.gershowrecycling.com.

Gesing, A., Shredder practice for preparation of nonmetallic concentrates and potential for particle sorting of these concentrates, presented at *SgS 2006*, Aachen, Germany, March 29, 2006. http://www.gesingconsultants.com/publications/Shredder-pract-sgs06-pres.pdf.

Gesing, A., Elemental analysis and chemical–composition–based material separation and blending, in *Light Metals 2007*, Sørlie, M. Ed., TMS-AIME, Warrendale, PA, 2007, p. 1099.

Gesing, A., Erdmann, T., and Gesing, M.A., Advanced industrial technologies for aluminium scrap sorting, presented at *Aluminium-21/Recycling*, St. Petersburg, Russia, 2010.

Gesing, A. and Harbeck, H., Profit motive, *Recycl. Today*, 47(10), 110, October 2009.

Graveman, D.F., *Replace Your Hammermill?* Magnatech Engineering, Tonganoxie, KS, January 30, 2009. http://www.magnatech-engineering.com/downloads/Replace-your-Hammermill.pdf.

Harding Metals, Inc., http://hardingmetals.com/company.htm.

Kessler Consulting, Inc., *Materials Recovery Facility Technology Review*, Kessler Consulting, Inc., Tampa, FL, September 2009. http://www.dep.state.fl.us/waste/quick_topics/publications/shw/recycling/InnovativeGrants/IGYear9/finalreport/Pinellas_IG8-06_Technology_Review.pdf.

Khullar, M., Surviving on scrap, *Scrap*, 66(5), 58, September/October 2009.

Komar Industries, Inc., Tiger Electro-Shear shredder, 12 August 2013 http://www.komarindustries.com/equipment/electro-shear-shredder.php.

Kvithyld, A., Engh, T.A., and Illés, R., Decoating of aluminium scrap in various atmospheres, in *Light Metals 2002*, Schneider, W., Ed., TMS-AIME, Warrendale, PA, 2002, p. 1055.

van Linden, J.H.L., Automotive aluminum recycling changes ahead, in *Light Metals 1994*, Mannweiler, U., Ed., TMS-AIME, Warrendale, PA, 1994, p. 1115.

Manouchehri, H.R., Looking at shredding plant configuration and its performance for developing shredding product stream, Jernkontorets Forskning, Stockholm, Sweden, D823, March 9, 2007. http://www.jernkontoret.se/ladda_hem_och_bestall/publikationer/stalforskning/rapporter/d_823.pdf.

Marshall, C., *Guillotine vs. Swing Beam Shear*, C. Marshall Fabrication Machinery, Inc., Simi Valley, CA, December 19, 2012. http://www.cmarshallfab.com/guillotine-vs-swing-beam-shear.

Maurice, D., Hawk, J.A., and Riley, W.D., Thermomechanical treatments for the separation of cast and wrought aluminum, in *4th International Symposium on Recycling of Metals Engineered Materials*, Stewart, D.L., Stephens, R., and Daley, J.C., Eds., TMS-AIME, Warrendale, PA, 2000, p. 1251.

McAvoy, B., McNeish, J., and Stevens, W., The Alcan decoater process for UBC decoating, in *2nd International Symposium on Recycling of Metals Engineered Materials*, van Linden, J.H.L., Stewart, D.L., and Sahai, Y., Eds., TMS-AIME, Warrendale, PA, 1990, p. 203.

Meier–Staude, M., Schlett, Z., Lungu, M., and Baltateanu, D., A new possibility in eddy–current separation, *Miner. Eng.*, 15, 287, 1991.

Metso Minerals, http://www.metsominerals.com/inetMinerals/MaTobox7.nsf/DocsByID/F9790232BD1B5BF442256B440047DC5B/$File/1243_RO_EN.pdf.

Metso Minerals, http://www.metsominerals.com/inetMineals/MaTobox7.nsf/DocsByID/277BAEB1DECCC2FD42256B4100451C96/$File/1229_RAS_EN.pdf.

Metso Texas Shredder, *Lindemann ZMF & ZM Metal Crushers*, 20 May 2009, http://www.metsomaterialstechnology.com/recycling/mm_recy.nsf/WebWID/WTB-090716-22575-61848/$File/RZ_imagebildung_Broschüre_ENG_END.pdf.

Midwest, *Dust–Buster for Shredding and Recycling*, Midwest Industrial Supply, Canton, OH, December 19, 2012. http://www.midwestind.com/products-services/equipment/process-dust-control-systems/foam-dust-suppression-dust-buster.html.

Milton, D., Advances in downstream physical separation, presented *at ISRI Operations Forum*, Phoenix, AZ, October 4, 2010. http://www.recyclingtoday.com/FileUploads/file/recycling/1101_RT_DownstreamSeparation_Presentation.pdf.

Minter, A., India's scrap struggle, *Scrap*, 63(5), 47, November/December 2006.

Minter, A., Brazil rising, *Scrap*, 67(3), 58, May/June 2010.

Minter, A., The industrial revolution, *Scrap*, 68(2), 142, March/April 2011.

Mutz, S., Pretz, T., and van Looy, E., Picking—An "old" process for separation of non-ferrous metals?, in *EPD Congress 2003*, Schlesinger, M.E., Ed., TMS-AIME, Warrendale, PA, 2003, p. 301.

Ness, H., The automobile scrap processing industry, in *Impact of Materials Substitution on the Recyclability of Automobiles*, ASME, New York, 1984, p. 39.

Nijkerk, A.A. and Dalmijn, W.L., *Handbook of Recycling Techniques*, Nijkerk Consultancy, The Hague, the Netherlands, 1998.

Novelli, L.R., Equipment focus: Briquetters, *Scrap*, 63(2), 169, 2006.

Ohio Baler Co., http://www.ohiobaler.com/alligatorshear.htm.

Otaki, M., Trend of sorting technology of aluminum material from end–of–life vehicle scrap, *J. Jpn. Inst. Light Metals*, 2009, 59, 612.

Phillips, M., Destiny by density, *Recycl. Today*, 34(12), 66, December 1996.

Pietsch, W., Briquetting of aluminum swarf for recycling, in *Light Metals 1993*, Das, S.K., Ed., TMS-AIME, Warrendale, PA, 1993, p. 1045.

Pretz, T. and Julius, J., Metal waste, Chapter 6 in *Waste: A Handbook for Management*, Letcher, T.M. and Vallero, D.A., Eds., Elsevier, Amsterdam, the Netherlands, 2011, p. 89.

Ridall, M., The next wave, *Recycl. Today*, 47(4), 128, April 2009.

Roc–Impact s.a.r.l., Impact crusher, 12 August 2013, http://www.crusher-rocimpact.com/crushers/impact-crusher.php.

Rochester Aluminum Smelting Canada Limited, *Scrap Purchasing*, Rochester Aluminum Smelting Canada Limited, Concord, Ontario, Canada, February 10, 2009. http://www.rochesteraluminum.com/purchasing.htm.

Rossel, H., Ollenschläger, I., and Pietruck, R., Recycling of post–consumer aluminium packaging, in *3rd International Symposium on Recycling of Metals*, ASM International Europe, Brussels, Belgium, 1997, p. 95.

Ruan, J. and Xu, Z., Approaches to improve separation efficiency of eddy current separation for recovering aluminum from waste toner cartridges, *Environ. Sci. Technol.*, 46, 6214, 2012.

SGM Magnetics, *Aluminum Separator Type Sandjet Data Sheet*, SGM Magnetics, Sarasota, FL, 2009. http://www.holgerhartmann.fi/sites/default/files/sandjet_english_09.pdf.

Steeman, A., *Recycling Packaging Material with an Aluminium Component*, January 7, 2013. http://bestinpackaging.com/2012/02/02/recycling-packaging-material-with-an-aluminium-component/.

Steinert, *Magnetic Drum MT*, December 17, 2003a. http://www.steinert.de/uploads/media/Magnetic-Drums.pdf.

Steinert, *Suspension Magnet UM/AM*, December 17, 2003b. http://www.steinertus.com/index.php?eID=tx_nawsecuredl&u=0&file=fileadmin/downloads/Suspension-Magnet-UM-AM.pdf&t=1357231010&hash=0acab9c30a27715a8ca2ab11d1dead7e5f475d44.

Steinert, *Induction Sorting System ISS®*, November 20, 2003c. http://www.steinertus.com/index.php?eID=tx_nawsecuredl&u=0&file=uploads/media/Induction-Sorting-System-ISS.pdf&t=1357584928&hash=6a64740ebab5e9a9b8b2822d06127e30d55a9df4.

Steinert, *X-Ray Sorting System XSS*, May 10, 2007. http://www.steinertus.com/index.php?eID=tx_nawsecuredl&u=0&file=uploads/media/X-Ray-Sorting-System.pdf&t=1357594823&hash=b2c8d02988ebc1db2b2bdb07f402760aaee0789f.

Steinert, *Non-Ferrous Metal Separator/NES Eddy Current Separator*, April 27, 2012. http://www.steinert.de/fileadmin/downloads/Non-Ferrous_Metal_Separator.pdf.

Stevens, W. and Tremblay, F., Fundamentals of UBC decoating/delacquering for efficient melting, in *Light Metals 1997*, Huglen, R., Ed., TMS-AIME, Warrendale, PA, 1997, p. 709.

Sullivan, J.P., Recycling scrap wire and cable: The state of the art, *Wire J. Int.*, 18(11), 36, November 1985.

Sweetman, A., Keen, C., Healy, J., Ball E., and Davy, C., Occupational exposure to dioxins at UK worksites, *Ann. Occup. Hyg.*, 48, 425, 2004.

Taylor, B., Thinking small, *Recycl. Today*, 48(10), 94, November 2010.

Toto, D., No lightweight, *Recycl. Today*, 50(1), S12, January 2012.

Walker Magnetics, *Eddy Current Separation Equipment*, 17 June 2011, http://www.walkermagnet.com/Collateral/Documents/English-US/Eddy Current Separation Equipment.pdf.

Wens, B., Schmalbein, N., Gillner, R., and Pretz, T., Recovery of non-ferrous metals: Potential and accessibility from municipal solid waste, presented at *IEMA Knowledge Exchange 2010*, January 27, 2010, Manchester, U.K.

Wilson, D.C., Araba, A.O., Chinwah, K., and Cheeseman, C.R. Building recycling rates through the informal sector, *Waste Manag.*, 29, 629, 2009.

Zöllner, A., Processing aluminium scrap profitably, *MBM*, (395), 35, November 2003.

6 Beneficiation Practice

The previous chapter introduced the technology used to beneficiate aluminum scrap. The processor of scrap material has several options to choose from for turning collected material into a product that can be sold to customers. The choice of options depends on the cost of different processing technologies, the added value that the scrap will have when improved to a particular grade, the demand for particular grades of scrap, and the impact of government regulations. This chapter will examine the flowsheets by which different sources of scrap aluminum are processed and the rationale behind the choice of particular technologies. This will be done using the same categories of *ore* described in Chapter 3—new scrap, UBCs, automotive and transportation scrap, electrical and electronic scrap, and municipal solid waste (MSW).

NEW SCRAP

The variety of types of new scrap and the variety of ways in which it is collected and distributed make it impossible to generalize about how it is processed. However, compared with old scrap, new scrap has several advantages that make it easier to upgrade:

- *New scrap is a known*: The purchaser of new scrap knows how much material is being purchased and which alloys it consists of. The impurities in the scrap—iron, plastic, other metals—are also known. This makes it easier to design processing schemes.
- *New scrap is usually cleaner*: Because it has never been put in service, new scrap has less dirt and moisture on it and is usually in better condition. This makes it easier to generate a clean product for sale to remelting facilities.
- *New scrap does not require disassembly*: New scrap is usually obtained as pieces of alloy and is less likely to include attachments, paint, and other materials that will require physical separation. This reduces processing costs.

New scrap comes in many conditions and purities, and most of the beneficiation techniques described in the previous chapter are applied to one or more of the grades. However, some are more commonly used than others. These include the following:

- *Hand sorting*: Because a load of mixed new scrap often includes a limited number of alloys, hand sorting often makes it possible to produce single-alloy scrap products. Even if this is not possible, hand sorting can help meet specifications for other scrap grades, by removing impurities. Often, a dealer's experience and knowledge of his suppliers is useful in hand sorting, because the appearance of a piece of new scrap—the shape of a punching, the type of scrapped casting or part—will be sufficient to identify the alloy.

- *Screening*: This is used both to remove undersize pieces of aluminum and to separate out dirt and other extraneous matter.
- *Shredding*: This is done primarily to reduce large pieces of scrap (such as stamping skeletons or cable) to a size suitable for charging to a furnace. However, it is often done to turnings as well, because smaller turnings are easier to briquette (Nijkerk and Dalmijn, 1998).
- *Briquetting*: Used primarily for turnings and borings. This is usually performed before degreasing. It helps to squeeze out some of the oil and grease, in addition to turning small particles into an equivalent larger piece of scrap (Nijkerk and Dalmijn, 1998).
- *Thermal decoating*: If the amount of grease and oil in a scrap load is less than 1%, the scrap can be charged directly to a remelting furnace without concern over excessive smoke or melt loss. However, higher levels will require degreasing, using one of the thermal decoating technologies described in the previous chapter. Degreasing is usually required for borings and turnings.

MSW

Of all the sorting technologies introduced in the previous chapter, hand sorting remains the most common method of recovering aluminum and other recyclables from MSW. Hand sorting is performed at one or more of several locations:

- *By the waste generator*: The different blue bag systems used by municipalities were described in Chapter 3. Using dual- or multiple-stream recycling results in a UBC product with improved quality, that is, less moisture and fewer impurities. However, it is claimed that single-stream recycling results in improved participation, and there is a higher overall collection rate. The recent trend has been toward single stream (Kessler Consulting, 2009b; Morawski, 2010; Schaffer, 2009).
- *By the primary collector*: Figure 3.3 shows a truck specifically designed for collecting curbside recyclables. The truck contains separate compartments for each type of recyclable. As the driver retrieves blue bags, he sorts the contents into the separate compartments. This type of vehicle is required for use with dual-stream recycling. For single-stream recycling, it has become apparent that sorting at an MRF is a preferred option (Kessler Consulting, 2009b).
- *At centralized MRFs*: Most MRFs still use some sort of hand sorting of incoming refuse to eliminate large and bulky items that would interfere with subsequent processing; this includes items such as cardboard boxes, plastic bags, and occasionally aluminum siding. However, the use of hand sorting to recover smaller items such as UBCs is increasingly less common (Kranert et al., 2004). MRFs are becoming increasingly automated, as will be described later.
- *At the dump*: The role of dumpsite scavengers in recycling was described in Chapter 3. Scavenging is an effective means of recovering metal from refuse, but is increasingly unacceptable from a societal perspective and is generally replaced as a region develops (Wilson et al., 2009).

The flowsheet used by MRFs for processing municipal refuse depends on the type of input, the cost of equipment and land, and the price for recyclables. Screening is commonly used to (1) separate oversize materials such as paper and cardboard from the rest of the garbage or (2) separate undersize materials such as glass shards and bottle caps from the bulk of the material (Kessler Consulting, 2009b; Kranert et al., 2004). Trommels are the most common type of screen. Most MRFs also feature magnetic separation in their process. Magnetic separators recover a ferrous product that can be sold to steel recyclers, and they purify the nonferrous fraction of iron that would otherwise be a contaminant. The use of eddy current separators is increasingly a standard method for generating an aluminum fraction with fewer impurities; models of ECS have been introduced that are effective enough to recover aseptic packaging as well as UBCs (see Chapter 3). Several other sensor-based sorting devices may also be useful for recovering aluminum from trash, including optical and induction sorting.

The importance of product purity in the aluminum recovered by MRF facilities is highlighted in Table 6.1. This is a set of specifications for *grades* of MRF-recovered aluminum, listing the maximum of various impurity elements allowed in each grade. (The specifications, produced by the American ASTM [1999], have since been abandoned.) Grades 5 and 6 are suitable for reuse in the production of

TABLE 6.1
Chemical Specification for Grades of Aluminum Scrap Produced by MRFs

	Composition (Max.% Allowable)					
Element	Grade 1	Grade 2	Grade 3	Grade 4	Grade 5	Grade 6
Silicon	0.30	0.30	0.50	1.00	9.00	9.00
Iron	0.60	0.70	1.00	1.00	0.80	1.00
Copper	0.25	0.40	1.00	2.00	3.00	4.00
Manganese	1.25	1.50	1.50	1.50	0.60	0.80
Magnesium	2.00	2.00	2.00	2.00	2.00	2.00
Chromium	0.05	0.10	0.30	0.30	0.30	0.30
Nickel	0.04	0.04	0.30	0.30	0.30	0.30
Zinc	0.25	0.25	1.00	2.00	1.00	3.00
Lead	0.02	0.04	0.30	0.50	0.10	0.25
Tin	0.02	0.04	0.30	0.30	0.10	0.25
Bismuth	0.02	0.04	0.30	0.30	0.10	0.25
Titanium	0.05	0.05	0.05	0.05	0.10	0.25
Others (each)	0.04	0.05	0.05	0.08	0.10	0.10
Others (total)	0.12	0.15	0.15	0.20	0.30	0.30
Aluminum	Bal.	Bal.	Bal.	Bal.	Bal.	Bal.

Source: ASTM (American Society for Testing and Materials), *Standard Specification for Municipal Aluminum Scrap (MAS)*, Standard E 753-80, Conshohocken, PA, 1999. With permission.

casting alloys; grades 1–4 are potentially useful for producing various types of wrought alloy. The high limits for magnesium content reflect the predominance of UBCs in the aluminum recovered by MRFs. The iron, chromium, and nickel limits reflect the challenge of removing stainless steel from aluminum. In addition to the chemical elements listed in Table 6.1, aluminum recovered by MRFs should have no more than 1 wt.-% fines for *Class A* product or 3% for lower-value Class B. A maximum of 2% loose combustible material is specified, as is a maximum of 0.5% moisture. More recent specifications by the American recycler Evermore Recycling (2012) also specify limits on fines and moisture content. MRFs have a difficult time meeting these standards, so it is not uncommon for bales of recovered UBCs to be sent for further upgrading before they can be accepted by a smelting facility (Schaffer, 2009). Income from the sale of UBCs is a substantial fraction of the income received by MRFs (Kessler Consulting, 2009a), so efforts are being made to improve sorting operations to avoid this extra step (Taylor, 2006).

UBCS AND OTHER PACKAGING

UBCs are among the few types of old aluminum scrap recycled primarily into wrought alloy rather than cast. As a result, the quality requirements for UBC scrap are tighter than for other grades. This means that more upgrading is performed on this than on other scrap types.

Scrap UBCs come primarily from two sources: cans recovered by blue bag recycling programs, MRFs, and scavengers from municipal garbage, and cans specifically collected by buy-back centers and similar operations. In either case, the collector will usually perform some upgrading of the collected cans before selling them to the remelting facility (Broughton, 1994; Harler, 1998; Steverson, 1995). Beneficiation of the cans by the collector often includes (1) flattening the collected cans, (2) magnetic separation of iron and steel, and (3) baling or logging prior to shipping. A significant concern during flattening and baling is the presence of aluminum aerosol cans, which contain fluids under pressure. These cans are not common, but the contents sometimes include flammable liquids and are potential explosion hazards (Smith et al., 2001). The level of hazard that aerosol cans pose is a matter of debate. Another problem in UBC scrap can be the presence of used hypodermic needles (Goodrich, 2001). The steel in these needles is nonmagnetic stainless, and so they are difficult to remove.

The scrap grade specifications in Figure 4.1 include several for UBCs (ISRI, 2012). Taldon (Figure 6.1) is the most important of these, since most UBCs are baled prior to shipping. The specification excludes any material other than UBCs. Other container materials such as plastic and steel are specifically unwelcome. Plastic clogs up the grates in shredders, and the burning of plastic in decoating furnaces causes loss of temperature control and damage to the furnace (Novelis, 2006).

One of the more surprising impurities that UBC processors deal with is lead (McNealy, 2008; Novelis, 2006). This appears to come mostly from the dirt attached to the cans as well as from deliberate contamination (Taylor, 2001b). Since even small levels of lead can injure the properties of 3xxx series alloys, processors and remelters go to considerable lengths to eliminate the dirt. Even so,

FIGURE 6.1 Baled old UBC scrap (Taldon). (Photo courtesy of Ray Peterson, Aleris International.)

remelted UBCs often contain 50–100 parts per million lead, and this is an ongoing concern.

Once the baled UBCs are received by a remelt facility, a series of additional upgrading processes are performed (Evans and Guest, 2000; Taylor, 2001b). The first is shredding of the bales, using a low-speed shredder of hammer mill. This (1) reduces the cans to a uniform particle size, (2) exposes the lacquered inside surface of the cans, making them easier to decoat, (3) allows trapped moisture or liquids to be drained away, and (4) releases trapped contaminants in the cans, allowing them to be removed. The shredded cans are then screened and conveyed under another magnetic separator.

Once the impurities are removed, the final step before remelting is decoating. The environmental and hygienic advantages of decoating were described in the previous chapter. Decoating is also useful because it preheats the scrap (Evans and Guest, 2000; Thornton et al., 2007). This reduces the time needed to melt the UBCs in the melting furnace. Because of this, more metal can be melted per hour in a furnace. In addition, utilizing the fuel value of the VOCs reduces total required energy costs. The presence of plastic and rubber in the UBCs makes temperature control difficult, so rotary decoating furnaces (which are less sensitive to temperature variations) are more commonly used.

Separate processing of foil and low-aluminum packaging is rare in North America, since separation of this material is rarely required or encouraged (Shoch and Gendell, 2011). In Europe, recycling foil packaging is often required. The depulping process described in the previous chapter is used (Kmeco and Fedorishin, 2010); as pointed out, this produces a polyethylene/aluminum composite residue that is difficult to separate. While pyrolysis is possible (see previous chapter), the cost is difficult to justify. Life-cycle analysis performed by Varžinskas et al. (2012) has shown that turning this composite into roof shingles may be the best option.

AUTOMOTIVE SCRAP

Figure 6.2 shows a *traditional* flowsheet for the recovery of aluminum from scrapped automobiles (Cui and Roven, 2010). This process has changed considerably over the years, in part due to the increasing importance of recovering the aluminum in the hulk. The development of new sorting technology described in the previous chapter will likely change this flowsheet more in the future.

As mentioned in Chapter 3, the first step in recycling an automobile is the dismantling process. This can be divided into three parts (Bebelaar and Dalmijn, 2007; Paul, 2007). The first is *pretreatment,* which is the removal of those items that are mandated either by law or shredder operators. These include all fluids, the batteries, tires, airbags, and mercury switches. Airbag and mercury switch removal is mandated by shredder operators, for safety and environmental reasons (Fischer, 2007a).

The second part of dismantling is removal of parts that can be reused or reconditioned. A long list of parts are candidates for reuse (Nakajima and Vanderburg, 2005; Paul, 2007), depending on the model of the vehicle; as described in Chapter 3, used cars (especially newer models) are often made available for parts recovery for some time before being further processed (Taylor, 2012). The third step of dismantling is the removal of parts whose value as recyclable material justifies the expense of specifically recovering them. The best example of this is catalytic converters, which contain platinum-group metals worth recovering. Aluminum parts that have enough potential value to be deliberately removed from a vehicle before shredding include

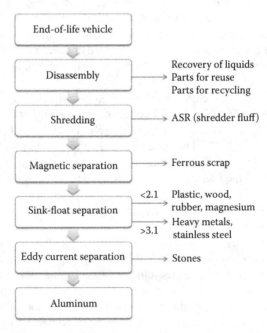

FIGURE 6.2 Generic flowsheet for processing of scrap automobiles. (After Cui, J. and Roven, H.J., *Trans. Nonferrous Met. Soc. China*, 20, 2057, 2010.)

wheels, compressors, alternators, water pumps, transmissions, and possibly engine blocks and heads. Dismantling is increasingly important as a recycling process, and designing cars that can be easily dismantled (without coming apart more easily on their own) is a priority of automakers (Fischer, 2007a). European automakers have in fact formed partnerships with automotive dismantlers to improve parts recovery and reuse (Nakajima and Vanderburg, 2005). Even so, the parts recovered from a typical dismantled auto contain only about 20% of the aluminum used in the vehicle.

Most dismantling is still done by hand, but there is increasing interest in assembly dismantling (Manouchehri, 2007; West–Japan Auto Recycle Co., Ltd., 2004). Figure 6.3 shows a Japanese auto dismantling flowsheet in which parts of the body are separated from the engine and drive train before shredding. *Demanufacturing* operations of this sort are encouraged by regulations in Europe and Japan that require the recovery or recycling of up to 95% the content of scrapped automobiles by 2015 (Gesing, 2005; Kanari et al., 2003). White goods (refrigerators, air conditioners, washers/dryers) are also partially dismantled prior to shredding (Karpel, 2006).

Once the desired dismantling has been done, the remaining auto hulk is shredded (Gesing, 2006). Nearly all automobiles are shredded using the horizontal hammer mill described in the previous chapter. Figure 11.9 is a map of the over 200 shredding facilities in the United States as of 2011; as might be suspected, the location of shredders is largely determined by population density. A similar number of shredders are located in Europe. The trend in the United States has turned toward smaller shredders (Fowler, 2012), which allow operators to match capacity to the size of the local market. Another trend is toward the use of moist rather than dry shredding (Manouchehri, 2007); wet shredding appears to be less popular. In addition to dismantled auto hulks, shredders are also fed *whole* older cars, white goods, metal recovered from construction and demolition activities, and other transportation scrap. Maintaining a consistent mix of these inputs is important to ensuring smooth shredder operation.

Figure 6.4, from the 2010 presentation by Gesing et al. (2010), describes the processing of dry shredding product. The *drum magnet* that recovers ferrous material is likely a two-drum arrangement, as shown in the previous chapter. If moist instead of dry shredding is used, the cyclone and baghouse receive material only from the Z-box air classifier and the product screens. The return of oversize material from the screens to the shredder is important because downstream devices such as eddy current and heavy media separators operate best with uniform size feed. The second magnetic separation helps eliminate ferrous material from the feed to the ECS. This prevents damage to the separator and produces a purer *NMMC* heavy-metal stream. The three nonmetallic streams are collectively known as *automotive shredder residue (ASR)* or *shredder fluff.* The product moving from the screens to the ECS is known as *nonmagnetic shredder fraction (NMSF)* and is equivalent to the Zorba scrap grade described in Figure 4.1. It contains most of the nonferrous metals in the auto hulk, along with rubber, glass, and some unrecovered plastic. The analysis of NMSF varies depending on what was input to the shredder, but ranges from 50% to over 90% aluminum. This value will increase as aluminum usage in automobiles continues to grow. About 81% of the aluminum input to the shredder winds up in the

Dismantling flow

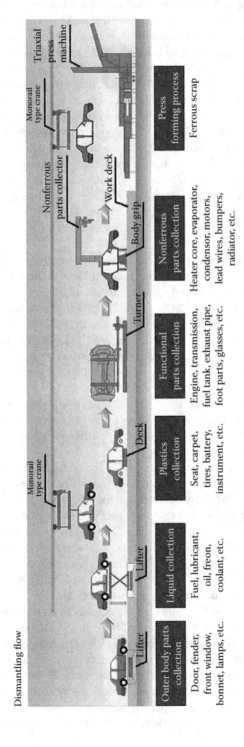

FIGURE 6.3 Auto dismantling flowsheet. (From West–Japan Auto Recycle Co., Ltd., Profile of West–Japan Auto Recycle Co., Ltd., August 12, 2004, http://www.env.go.jp/earth/coop/neac/neac12/pdf/04l.pdf. With permission.)

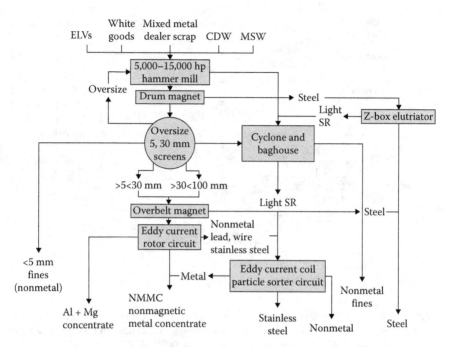

FIGURE 6.4 Sorting flowsheet for auto shredder product. ECS-based sorting of nonferrous concentrate (Zorba) from automobile shredding. (From Gesing, A. et al., Advanced industrial technologies for aluminium scrap sorting, presented at *Aluminium-21/Recycling*, St. Petersburg, Russia, 2010. With permission.)

NMSF and another 18% in the shredder fluff (Manouchehri, 2007). (The remainder is lost in the ferrous product.)

Although Figure 6.4 shows the use of ECS for sorting NMSF, there are several process options. One that continues to be popular is the use of hand sorting. Hand sorting is a particularly effective technology for dealing with the oversize material from the shredder, since it eliminates the cost of reshredding (Manouchehri, 2007). As a result, an active market exists for oversize NMSF and Zorba, although increasing labor costs are beginning to have an impact.

If NMSF is kept for further downstream processing, the next step is one of two options. Figure 6.5 shows the older of the two, which starts with HMS (Cui and Roven, 2010; Gesing et al., 2010; Seebacher et al., 2006). In practice, the feed to HMS is often washed first in a rising-current water separator such as that described in the previous chapter (Manouchehri, 2007); this removes much of the plastic and remaining shredder fluff from the NMSF, improving the efficiency of the HMS units. The figure shows two-stage HMS, with fluids maintained at specific gravities of 2.0 and 3.5. The floated material from the first stage is sometimes known by the ISRI specification Zeppelin (ISRI, 2012) and contains most of the nonmagnetic material in the feed, along with magnesium and some of the aluminum (hollow and thin-gauge pieces have a lower apparent density). Figure 6.6 shows the floated material from the second stage, which contains the rest of the aluminum as well as

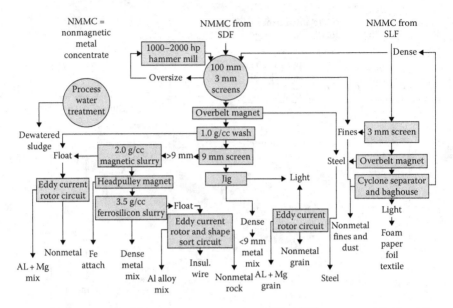

FIGURE 6.5 Flowsheet for HMS-based processing of NMMC from automobile shredding. (From Gesing, A. et al., Advanced industrial technologies for aluminium scrap sorting, presented at *Aluminium-21/Recycling*, St. Petersburg, Russia, 2010. With permission.)

FIGURE 6.6 Upgraded auto shred (Twitch). (Photo courtesy of Ray Peterson, Aleris International.)

minor amounts of insulated wire and denser nonmetallic material (stones, glass). It is marketed using the ISRI specification Twitch (see Figure 4.1). The heavy fraction from the second HMS stage contains higher-density metals such as copper, zinc, and stainless steel and is known by the ISRI specification Zebra.

The Twitch recovered from the second HMS can be further upgraded by processing in an eddy current separator (Gesing et al., 2010; Manouchehri, 2007). This separates the aluminum from the wire and nonmetallics. The Zeppelin recovered from the first separator can also be upgraded using ECS, but will have to be further processed to separate the aluminum from the magnesium, using either hand picking or a sensor-based technique (Seebacher et al., 2006).

The disadvantages of using HMS as the first step in sorting NMSF include incomplete separation of aluminum from magnesium and the difficulty of separating aluminum from insulated wire. As the use of electronics in automobiles grows, this becomes a more significant problem. As a result, the flowsheet in Figure 6.4 shows the use of ECS as the first sorting step instead (Gesing et al., 2010). ECS generates three products: a nonmetallic stream containing the heaviest metals (lead, stainless steel), along with most of the rubber and plastic; *NMMC* containing high-density metals (copper, zinc) and pieces of aluminum attached to other metals; and a light-metal stream containing magnesium and the rest of the aluminum. The light-metal stream resembles the Zeppelin product from HMS and will require further upgrading. ECS does a better job than HMS at separating the aluminum in NMSF from the wire and eliminates the other concerns of using HMS (see Chapter 5).

Figure 6.7 shows what may be the future of automotive recycling (Fischer, 2007b). The system relies on ECS as the first sorting tool for handling NMSF as before (Gesing et al., 2010), but now includes a second separator specifically for processing fines. The ECS units produce only two streams: a *nonferrous* stream containing the metallic content and a nonmetallic fluff. Aluminum is recovered from the nonferrous stream using the x-ray transmission system described in the previous chapter, most likely a DEXRT to eliminate size and shape effects. The flowsheet also relies on x-ray fluorescence to process heavier nonferrous metals and to produce a copper-free ferrous scrap product. Future NMSF processing may also utilize induction sorting to recover stainless steel, color sorting to separate out shredded circuit boards, and LIBS to separate aluminum and magnesium and to sort pieces of aluminum alloy by series.

The aluminum content of shredder fluff is only 1.5%–2%, and the percentage of other metals is even smaller (Srogi, 2007). As a result, shredder fluff has traditionally been landfilled or burned for its fuel value. However, this is increasingly unacceptable from an environmental standpoint. Regulations such as the European Union recycling targets discussed in Chapter 3 have made it increasingly necessary to process shredder fluff (Kanari et al., 2003). The increasing value of the metal in shredder fluff (particularly the copper) has also spurred efforts to process this stream for metal recovery. ECS units designed for treating fines can be used to recover metal from shredder fluff; induction sorters are also effective (Fischer, 2007b; Gesing, 2006). As previously mentioned, 18% of the aluminum input to the shredding process winds up in the fluff (Manouchehri, 2007), so an efficient process for metals recovery from this stream could be cost effective.

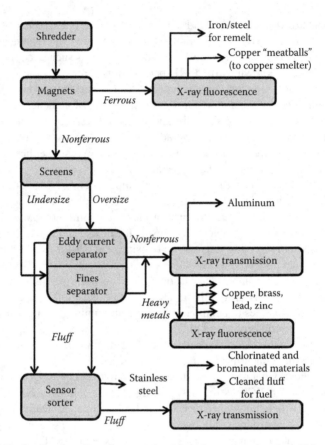

FIGURE 6.7 Sorting flowsheet for auto shred based on sensor sorting. (After Fischer, T., *Scrap,* 64(6), 113, 2007b.)

ELECTRICAL AND ELECTRONIC SCRAP (SULLIVAN, 1985)

Upgrading schemes for wire and cable scrap vary according to the type of scrap being processed. As a result, there is no typical flowsheet for the process. However, several unit operations are common to most cable scrap processing sequences.

The first step in cable processing is often hand sorting, which separates cable scrap by metal (aluminum or copper) and by size (Muchová and Eder, 2010). This is typically followed by a shearing step, meant to reduce the cable to 25–50 cm lengths. Portable alligator or guillotine shears are commonly used for this. Some operations now use high-speed shredders instead and *prechop* to lengths as small as 2 cm (Taylor, 2001a). Large pieces are also removed by hand prior to chopping (Fowler, 2008).

Figure 6.8 shows the construction of a typical underground cable (Nijkerk and Dalmijn, 1998). Recycling this cable is more difficult than recycling the typical plastic-coated above-ground cable. In addition to the materials shown here, underground cables can also have lead sheathing, mastic, oil, grease, and tar-impregnated paper.

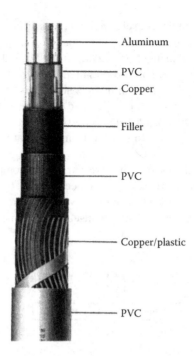

FIGURE 6.8 Construction of an underground cable. (From Nijkerk, A.A. and Dalmijn, W.L., *Handbook of Recycling Techniques,* Nijkerk Consultancy, The Hague, the Netherlands, 1998. With permission.)

Because of this, wire-chopping scrap underground cable can present a serious safety and environmental hazard (Fowler, 2008). To mitigate this hazard, sheared underground cable is often stripped to separate the coating layers. The stripping opens up the various layers, and the metal content is pulled out by hand. The process is slow and justifiable only for large-diameter cables.

For other cables, chopping (also known as granulation) follows shearing (Muchová and Eder, 2010). Chopping is a specialized form of low-speed shredding, intended to separate the metal in the cable from the plastic or rubber coating. The smaller the size of *chop* produced by this step, the greater the liberation of the metal from the insulation. However, chopping to a smaller size increases costs and reduces the productivity of the granulator. Some operations have a two-stage chopping sequence to minimize this bottleneck (Fowler, 2008). Final chop size is usually less than 1 cm.

The next step is often magnetic separation, to remove any steel pieces from the chop. Larger pieces of steel are usually removed by hand shearing prior to chopping, but small pieces may remain and must be removed. This is followed by separation of the metal from the insulation, using the difference between the specific gravity of the two. This is most commonly done using air tables, but air knives are also used, as are water-based rising current separators (Fowler, 2008). If the cable has not been sorted before chopping, a second classification step is needed to separate copper from aluminum. Properly chopped and sorted aluminum wire is marketed as ISRI grade Tall (ISRI, 2012).

The recovery of aluminum from electronic scrap is different than from other types. The reason is that the average aluminum content of discarded electronic equipment is small, about 4%–5% (Muchová and Eder, 2010). The value of other materials in the scrap, in particular copper and precious metals, is greater than that of the aluminum. As a result, recovery of the aluminum is not a high-priority concern. However, the recycling of scrap computers and associated equipment is increasingly mandated by law (Herat and Pariatamby, 2012). The development of recycling schemes that minimize the amount of waste to dispose of means that recovery of aluminum is important, in spite of its low value.

Electronic equipment has such a wide variety of compositions and constructions that a single generic recycling flowsheet cannot be constructed. However, Figure 6.9 illustrates a flowsheet including several common unit processes (Kumar et al., 2010). The first step is dismantling, usually performed by hand. Dismantling

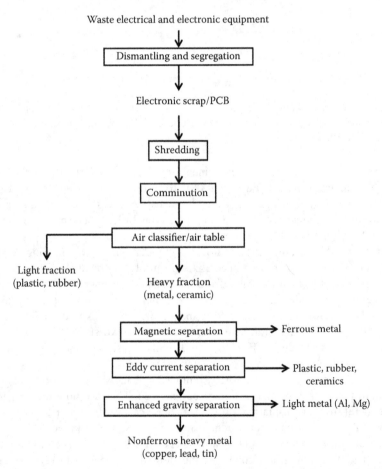

FIGURE 6.9 Generic flowsheet for processing of electronic scrap. (After Kumar, V., et al., *Korea Earth Syst. Sci. Eng.*, 47, 593, 2010.)

improves recovery of casings and structural parts, while minimizing contamination; since many of these parts are aluminum, it has particular importance here. It also increases the possibility for closed-loop recycling. However, dismantling is also labor intensive, and the parts that it recovers cannot be resold as in auto recycling. As electronic devices get smaller, the amount of metal to be recovered from dismantling declines as well (Stoklosa, 2011). As a result, electronics recyclers sometimes limit dismantling to recovery of materials that need to be recycled separately (batteries) or contain potentially hazardous components (lead, mercury).

Both hammer mills and knife shredders are used for comminution of electronic scrap. If disassembly is not performed, a hammer mill will be needed to reduce the metal content in the scrap. In some cases, a two-stage shredding system is used (Kang and Schoenung, 2005). The goal is to minimize the generation of fines while at the same time liberating the plastic in the e-waste from the metal content. Air classification separates the metal from the plastic, and ECS further purifies the metal stream. HMS completes the process of separating a light-metal stream from the heavier metals (in particular copper). ISRI scrap grade EM-2 describes electronic scrap that has been processed to produce a predominantly aluminum nonferrous product (ISRI, 2012). Optical sorting may also play a role in the downstream sorting of shredded e-waste (Milton, 2010).

RECOMMENDED READING

Gesing, A., Shredder practice for preparation of nonmetallic concentrates and potential for particle sorting of these concentrates, presented at *SgS 2006*, Aachen, Germany, March 29, 2006. http://www.gesingconsultants.com/publications/Shredder-pract-sgs06-pres.pdf.
Manouchehri, H.R., Looking at shredding plant configuration and its performance for developing shredding product stream, Jernkontorets Forskning, Stockholm, Sweden, D823, March 9, 2007. http://www.jernkontoret.se/ladda_hem_och_bestall/publikationer/stalforskning/rapporter/d_823.pdf.

REFERENCES

ASTM (American Society for Testing and Materials), *Standard Specification for Municipal Aluminum Scrap (MAS),* Standard E 753–80, Conshohocken, PA, 1999.
Bebelaar, D. and Dalmijn, W.L., Latest developments in car recycling in the Netherlands, in *Light Metals 2007,* Sørlie, M., Ed., TMS-AIME, Warrendale, PA, 2007, p. 1089
Broughton, A.C., A tradition of recycling, *Recycl. Today,* 32(10), 36, October 1994.
Cui, J. and Roven, H.J., Recycling of automotive aluminum, *Trans. Nonferrous Met. Soc. China,* 20, 2057, 2010.
Evans, R. and Guest, G., *The Aluminium Decoating Handbook,* Stein Atkinson Stordy Ltd., Wolverhampton, U.K., June 2000. http://infohouse.p2ric.org/ref/26/25275.pdf.
Evermore Recycling, *UBC Product Quality Specification,* September 6, 2012. http://www.alcoa.com/rigid_packaging/en/pdf/EVMR-089%20Rev%2008%20UBC%20Product%20Quality%20Specifications%20updated.pdf.
Fischer, T., The design for dismantling dilemma, *Scrap,* 64(5), 72, September/October 2007a.
Fischer, T., Equipment focus: Downstream separation systems, *Scrap,* 64(6), 113, November/December 2007b.
Fowler, J., Equipment focus: Compact wire processors, *Scrap,* 65(2), 231, March/April 2008.
Fowler, J., Equipment focus: Smaller auto shredders, *Scrap,* 69(6), 125, November/December 2012.

Gesing, A., Shredder practice for preparation of nonmetallic concentrates and potential for particle sorting of these concentrates, presented at *SgS 2006*, Aachen, Germany, March 29, 2006. http://www.gesingconsultants.com/publications/Shredder-pract-sgs06-pres.pdf.

Gesing, A., Erdmann, T., and Gesing, M.A., Advanced industrial technologies for aluminium scrap sorting, presented at *Aluminium-21/Recycling*, St. Petersburg, Russia, 2010.

Gesing, A.J., Aluminium, recycling and transportation, presented at *Aluminium 2005*, Kliczkow Castle, Poland, October 13, 2005. http://www.gesingconsultants.com/publications/50926.pdf.

Goodrich, M., Leaders of the pack, *Recycl. Today*, 39(5), 64, May 2001. Harler, C., Benchmark your UBC operation against the best, *Recycl. Today*, 36(9), 69, September 1998.

Herat, S. and Pariatamby, A., E-waste: A problem or an opportunity? Review of issues, challenges and solutions in Asian countries, *Waste Manag. Res.*, 30, 1113, 2012.

ISRI (Institute of Scrap Recycling Industries), *Scrap Specifications Circular 2012*, October 8, 2012. http://www.isri.org/iMIS15_Prod/ISRI/_Program_and_Services/Commodities/Scrap_Specifications/ISRI/_Program_and_Services/Scrap_Specifications_Circular.aspx?hkey=5c76eb15-ec00-480e-b57f-e56ce1ccfab5.

Kanari, N., Pineaul, J.-L., and Shallari, S., End-of-life vehicle recycling in the European Union, *JOM,* 55(8), 15, August 2003.

Kang, H.-Y. and Schoenung, J.M., Electronic waste recycling: A review of U.S. infrastructure and technology options, *Res. Conserv. Recyc.*, 45, 368, 2005.

Karpel, S., Recycling Japan, *Metal Bull. Monthly,* (424), 33, April 2006.

Kessler Consulting, *Materials Recovery Facility Feasibility Study,* September 2009a. http://www.dep.state.fl.us/waste/quick_topics/publications/shw/recycling/InnovativeGrants/IGYear9/finalreport/Pinellas_IG8–06_MRF_Feasibility_Study.pdf.

Kessler Consulting, *Materials Recovery Facility Technology Review,* September 2009b. http://www.dep.state.fl.us/waste/quick_topics/publications/shw/recycling/InnovativeGrants/IGYear9/finalreport/Pinellas_IG8–06_Technology_Review.pdf.

Kmeco, R. and Fedorishin, V., Drink packaging processing—A hot issue of today, *Pulp and Paper International,* 52(6), 25, June 2010.

Kranert, M., Schultheis, A., and Steinbach D., *Methodology for Evaluating the Overall Efficiency of Sorting in Europe, Comparable Data and Suggestion for Improving Process Efficiency Analysed within the Project,* AWAST Deliverable, February 10, 2004. http://www.dep.state.fl.us/waste/quick_topics/publications/shw/recycling/InnovativeGrants/IGYear9/finalreport/Pinellas_IG8–06_Technology_Review.pdf.

Kumar, V., Lee, J.-C., Jeong, J., and Kim, B.-S., Review on mechanical recycling of end-of-life electrical and electronic equipments for recovery of metallic components, *Korea Earth Syst. Sci. Eng.*, 47, 593, 2010.

Manouchehri, H.R., *Looking at Shredding Plant Configuration and Its Performance for Developing Shredding Product Stream*, Jernkontorets Forskning, Stockholm, Sweden, D823, March 9, 2007. http://www.jernkontoret.se/ladda_hem_och_bestall/publikationer/stalforskning/rapporter/d_823.pdf.

McNealy, P.R., *Section 3: Contaminants: What They Are and Why They're an Issue,* May 19, 2008. http://www.alcoa.com/rigid_packaging/en/pdf/quality_rating_procedures.pdf.

Milton, D., Advances in downstream physical separation, presented at *ISRI Operations Forum*, Phoenix, AZ, October 4, 2010. http://www.recyclingtoday.com/FileUploads/file/recycling/1101_RT_DownstreamSeparation_Presentation.pdf.

Morawski, C., Single–stream uncovered, *Resour. Recycl.,* 29(2), February 21, 2010. http://www.container-recycling.org/assets/pdfs/media/2010–2-SingleStreamUncovered.pdf.

Muchová, L. and Eder, P., End–of–Waste criteria for aluminium and aluminium alloy scrap: Technical proposals, JRC Scientific and Technical Reports, EUR 24396, 2010. http://ftp.jrc.es/EURdoc/JRC58527.pdf.

Nakajima, N. and Vanderburg, W.H., A failing grade for the German end–of–life vehicles take–back system, *Bull. Sci. Technol. Soc.*, 25, 170, 2005. http://bst.sagepub.com/content/25/2/170.full.pdf.

Nijkerk, A.A. and Dalmijn, W.L., *Handbook of Recycling Techniques,* Nijkerk Consultancy, The Hague, the Netherlands, 1998.

Novelis, *UBC Specifications and Guidelines,* Novelis, São Paulo, Brazil, July 2006. http://www.benefits.us.novelis.com/NR/rdonlyres/72030193-07EB-4408-A5ED-FA28D-5CE8EA4/5536/UBCSpecificationsandGuidelinesforNovelisNorthAmeri.pdf.

Paul, R.T., The success of vehicle recycling in North America, in *Light Metals 2007,* Sørlie, M., Ed., TMS-AIME, Warrendale, PA, 2007, p. 1115.

Schaffer, P., Why UBC consumers are finally warming up to MRFs, *Am. Metal Market Monthly,* 118(2), 26, March 2009.

Seebacher, H., Sunk, W., Antrekowitsch, H., and Klade, M., Recycling von aluminium in der automobilindustrie, *Aluminium,* 82(1/2), 24, January/February 2006.

Shoch, E. and Gendell, A., *Closing the Loop: Design for Recovery Guidelines: Aluminum Packaging,* Green Blue Institute, Charlottesville, VA, June 30, 2011. http://www.greenblue.org/wp-content/uploads/2011/07/CTL_Design-for-Recovery_Aluminum_web.pdf.

Srogi, K., An overview of current processes for the thermomechanical treatment of automobile shredder residue, *Clean Technol. Environ. Policy,* 10, 235, 2007.

Steverson, W.B., Can sheet performance as a function of UBC quality, in *Aluminum Alloys for Packaging III,* Das, S.K., Ed., TMS-AIME, Warrendale, PA, 1998, p. 151.

Stoklosa, K., Process improvement, *Recycl. Today,* 49(8), 60, August 2011.

Sullivan, J.F., Recycling scrap wire and cable: The state of the art, *Wire J. Int.,* 18(11), 36, November 1985.

Taylor, B., Prep work, *Recycl. Today,* 39(10), 18, October 2001a.

Taylor, B., The fiery UBC furnace, *Recycl. Today,* 39(12), 18, December 2001b.

Taylor, B., Ship shape, *Recycl. Today,* 44(8), 56, August 2006.

Taylor, B., In–house service, *Recycl. Today,* 50(9), 44, September 2012. http://www.recyclingtoday.com/rt0912-u-pull-it-profile.aspx.

Thornton, T., Hammond, C., van Linden, J., Campbell, P., and Vild, C., Improved UBC through advanced processing, in *Light Metals 2007,* Sørlie, M., Ed., TMS-AIME, Warrendale, PA, 2007, p. 1191.

Varžinskas, V., Stanisklas, J.K., and Knasyte, M., Decision-making support system based on LCA for aseptic packaging recycling, *Waste Manag. Res.,* 30, 931, 2012.

West–Japan Auto Recycle Co., Ltd., Profile of West–Japan Auto Recycle Co., Ltd., August 12, 2004. http://www.env.go.jp/earth/coop/neac/neac12/pdf/041.pdf.

Wilson, D.C., Araba, A.O., Chinwah, K., and Cheeseman, C.R., Building recycling rates through the informal sector, *Waste Manag.,* 29, 629, 2009.

7 Melting Furnace Fundamentals

Most of the scrap aluminum recycled today is remelted in a furnace fired with fossil fuels, most commonly natural gas. Furnaces of this type have been used since the recycling of aluminum began, and while the optimal design has changed, the fundamentals behind the melting of aluminum have not. Appreciation of these fundamentals is important to understanding the reasons for the numerous furnace designs and equipment innovations that have taken place over the past 20 years.

HEAT-TRANSFER KINETICS

Figure 7.1 presents a simple view of a typical *reverberatory* (*reverb*) furnace for melting solid aluminum. The furnace is typically fired with natural gas, causing heat to be transferred from the burner to the mixture of solid and/or molten aluminum in the *hearth* (furnace bottom) below. While much of the heat is transferred directly from the flame to the metal, some is transferred to the metal by indirect transfer from the flame to the refractory walls. The heat reverberates (bounces) off the walls and impinges on the aluminum in the hearth. Heat may bounce back and forth between the metal and the walls several times before it is absorbed by the metal, since aluminum reflects heat better than most metals. As a result, reverb furnaces feature shallow baths, maximizing the surface area per ton of contained metal.

Heat is transferred from the flame to the metal by one of two means: *radiation* and *convection*, the direct contact of hot gas on the surface of the bath itself. Heat transfer by radiation is governed by the following expression (Jenkins, 2000):

$$q_s = \varepsilon \sigma A \left(T_g^4 - T_s^4 \right) \tag{7.1}$$

where
- q_s is the rate of heat transfer, in watts per square meter of bath surface
- A is the amount of the metal surface exposed per square meter of bath surface (one if the metal is molten, less if not)
- σ is Planck's constant ($5.67 \cdot 10^{-8}$ W/m² K⁴)
- ε is the emissivity of the aluminum in the hearth
- T_g and T_s are the temperatures of the flame and the solid aluminum, respectively

ε is extremely low for polished aluminum (~0.05); as the metal heats up and acquires a thicker oxide surface, ε increases to about 0.25 and higher for alloys

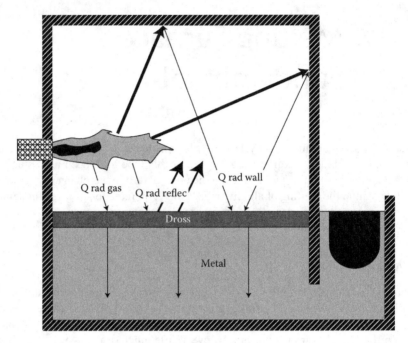

FIGURE 7.1 Cross section of reverberatory melting furnace, illustrating radiation heat transfer. (From Alchalabi, R. et al., Furnace operation of optimization via enhanced bath circulation, in *Light Metals 2002*, Schneider, W., Ed., TMS-AIME, Warrendale, PA, p. 739, 2002. With permission.)

(Chandler and Shull, 2009). A is increased by a thicker layer of solid scrap in the hearth and decreased by greater *porosity* (open space) in the scrap layer and by larger pieces of scrap in the charge. Equation 7.1 calculates only direct radiation heat-transfer rates, and not radiative heat transfer by reverberation. Equation 7.1 is also accurate only for surfaces directly facing the radiation source. For the surface not directly facing the radiation source, a *view factor* is used to calculate a reduced value of q_s.

Figure 7.2, from the work of Furu et al. (2010), illustrates how higher values of ε can reduce the time required to heat solid aluminum. Another strategy for increasing radiation heat transfer is increasing the freeboard above the melt (Lucas, 2012); increasing the thickness of the gas *blanket* in the furnace in turn increases reverberation off the walls and roof.

Because q_s is a function of T_g^4, raising the flame temperature is especially important in increasing heat-transfer rates by radiation. However, this has its drawbacks. Because surfaces directly facing the radiation source receive the most radiation, the metal nearest the flame heats much more rapidly than metal farther away from the flame. This causes overheating, which leads to excessive oxidation of the overheated metal. As a result, melt loss and dross formation are increased (Yap, 1995).

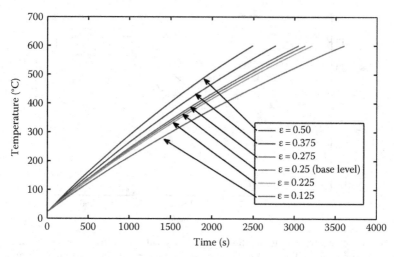

FIGURE 7.2 Heating rate of aluminum ingot surface as a function of surface emissivity. (From Furu, J. et al., Heating and melting of single Al ingots in an aluminum melting furnace, in *Light Metals 2010*, Johnson, J.A., Ed., TMS-AIME, Warrendale, PA, p. 679, 2010. With permission.)

The rate of heat transfer by convection is also calculated by a simple expression:

$$q_s = h \ (T_g - T_s) \tag{7.2}$$

where

q_s is again the rate of heat transfer per unit area of bath surface

T_g and T_s are the temperatures of the gas above the surface and of the solid (or liquid) metals, respectively

h is a convection heat transfer coefficient, determined by the composition, pressure, temperature, and flow characteristics of the gas; typical values for melting conditions are 10–30 W/m²-°C (Jenkins, 2000)

Higher pressures increase h, as does directing gas flow at the metal rather than parallel to it (Lucas, 2012). Convection heat transfer depends less on flame temperature than radiation but is still promoted by higher temperatures. Because no single point source is involved, it is less likely to cause overheating than radiation. However, it is also less effective at heat transfer; most transfer in industrial furnaces occurs by radiation (Buchholz and Rødseth, 2011; Furu et al., 2012). In addition, the surface turbulence caused by the gas can break through dross layers, exposing fresh metal and increasing melt loss (Yap, 1995).

Figure 7.3, from the results of Furu et al. (2010), illustrates heat flux rates by convection and radiation during natural gas heating of a solid aluminum surface. As time proceeds and the metal surface gets hotter, both rates decline; the decline of convection heat flux is linear, as predicted by Equation 7.2. The decline of radiation heat flux is more complex, partially because of the changing value of ε as the

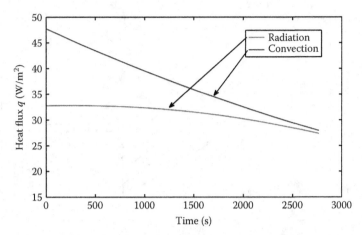

FIGURE 7.3 Heat transfer to an aluminum ingot surface by convection and radiation as a function of time. (From Furu, J. et al., Heating and melting of single Al ingots in an aluminum melting furnace, in *Light Metals 2010*, Johnson, J.A., Ed., TMS-AIME, Warrendale, PA, 2010, p. 679. With permission.)

surface heats up. The relative importance of radiation as a heat-transfer mechanism increases with metal temperature (Lucas, 2012), suggesting that optimal burner operating conditions should change during the melting process.

Once heat is transferred to the surface of the metal, it is transferred away from the surface to the bulk of the aluminum. In molten metal, this is done partly by convection; in solid and molten metal, conduction plays a large role as well. The expression governing heat transfer by conduction per unit area of melt surface is

$$q_c = \left(\frac{k}{L}\right)\left(T_{sur} - T_{bulk}\right) \tag{7.3}$$

where
  k is the thermal conductivity of the metal
  L the depth of the melt (or thickness of the solid piece)
  T_{sur} the temperature at the surface
  T_{bulk} the temperature of the bulk of the melt (or solid charge)

The thermal conductivity of solid aluminum is twice that of liquid (Lucas, 2012), but this is reduced in scrap loads by the gaps between the charged pieces. (Aluminum alloys are less conductive than pure aluminum as well.) Aluminum has a high thermal conductivity but that of dross is lower, and heat transfer through the dross layer can be a limiting factor in melting rates (Alchalabi et al., 2002; Lucas, 2012).

Because heat is constantly being transferred to the metal surface in a melting furnace, the surface temperature will always be higher than that of the bulk. If conduction is the primary heat-transfer mechanism, the *stratification* in the molten metal that results can be severe (Alchalabi et al., 2002; Matsuzaki et al., 2011; Zhou et al.,

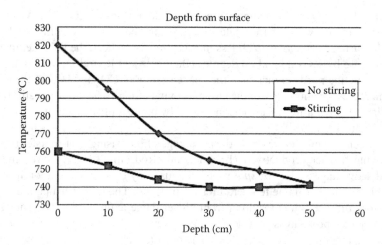

FIGURE 7.4 Impact of stirring on temperature vs. melt depth in molten aluminum. (From Matsuzaki, K. et al., Evaluation of effects of stirring in a melting furnace for aluminum, in *Light Metals 2011*, Lindsay, S.J., Ed., TMS-AIME, Warrendale, PA, p. 1199, 2011. With permission.)

2010), with temperature differences between the top and bottom of the melt of 100°C or higher. As a result, increasing the convective heat transfer coefficient h is important. Since the temperature and composition of the metal are fixed, stirring is the most effective method of increasing h. Figure 7.4 shows the impact of stirring on temperature stratification in an aluminum melt; stirring also helps reduce composition stratification, which can occur when higher-density alloying elements *settle* to the bottom of the melt. Several stirring devices are described in the next chapter with this in mind.

COMBUSTION

Most reverb furnaces burn natural gas to provide energy for melting. Natural gas consists mostly of *methane* (CH_4), with some *ethane* (C_2H_6). Several other hydrocarbons and other gases are present in smaller concentrations; the composition varies from place to place. The overall combustion of natural gas reacts the hydrocarbons with oxygen to generate carbon dioxide and water vapor:

$$CH_4 + 2O_2 = CO_2 + 2H_2O \text{ (g)} \tag{7.4}$$

The energy produced by this reaction heats the product gases to a characteristic flame temperature. The radiation from this flame and the heat transfer from the product gases to the metal by convection heat up and melt the scrap charge in the furnace.

Several characteristics of the flame are adjustable. The most important is the flame temperature. Normal air contains roughly 21% oxygen and 79% nitrogen. Nitrogen does not participate in the combustion reaction, but it does absorb much of the heat that combustion generates. As a result, with more gas to absorb the same amount of energy, flame temperatures are relatively low (<2000°C). When pure oxygen is mixed with air to create *enriched* air, the amount of nitrogen brought into the furnace decreases. With less nitrogen to adsorb heat, the flame temperature can increase by several hundred degrees (Buchholz and Rødseth, 2011). In addition, eliminating the nitrogen increases the concentration of CO_2 and H_2O in the off-gas; this increases the emissivity of the gas, improving radiative heat transfer still further. The flame temperature can also be increased by preheating the incoming air (Lucas, 2012; Zhou et al., 2010).

High flame temperatures have the advantage of increasing heat-transfer rates and improving furnace productivity (Buchholz and Rødseth, 2011). However, there are also disadvantages. The *hot spot* problem caused by excessive radiation heating is a problem at higher flame temperatures (Yap, 1995). The higher temperature of the refractory furnace lining resulting from higher flame temperatures also shortens its life. A bigger problem is unwelcome side reactions that are more likely to occur at flame temperatures above 2000°C. The most troublesome are reactions between nitrogen and unreacted oxygen to form nitrogen oxides (Li et al., 2006):

$$N_2 + O_2 = 2NO \tag{7.5}$$

$$N_2 + 2O_2 = 2NO_2 \tag{7.6}$$

NO and NO_2 are the main constituents of NO_x, a significant pollution concern. Reactions 7.5 and 7.6 become more feasible as the flame temperature increases. However, lower partial pressures of N_2 in the exhaust gas caused by the use of oxygen-rich air act to reduce NO_x generation. As the gas temperature decreases, NO and NO_2 will decompose to N_2 and O_2 if given enough time. *Staged* combustion of natural gas helps to avoid hot spots and NO_x formation (Nieckele et al., 2011).

If the amount of oxygen mixed with the fuel in the burner is not sufficient to immediately burn the fuel, reaction 7.4 will happen by stages. The first stage is *pyrolysis* of the fuel to generate hydrogen and small particles of soot (Wechsler and Gitman, 1990):

$$CH_4 = C + 2H_2 \tag{7.7}$$

The soot and hydrogen then react with more oxygen to produce carbon monoxide and water vapor:

$$C + 2H_2 + \frac{3}{2}O_2 = CO + 2H_2O \tag{7.8}$$

The carbon monoxide then reacts with more O_2 to combust completely to CO_2. This reaction sequence is significant because the soot increases the *luminosity* of the flame (Solovjov and Webb, 2005), that is, its ability to radiate. Sootier flames tend to be more spread out and radiate from a larger area. On the other hand, the slower reaction sequence tends to generate a lower flame temperature.

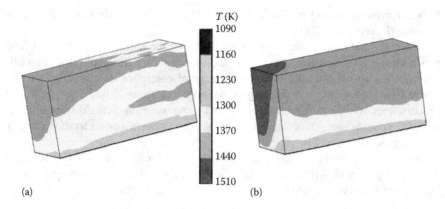

T (K)
1090
1160
1230
1300
1370
1440
1510

(a)   (b)

FIGURE 7.5 CFD simulation of flame temperatures produced by burning (a) natural gas and (b) liquid fuel oil in a reverberatory furnace. (From Nieckele, A.O. et al., *Appl. Therm. Eng.*, 31, 841, 2011. With permission.)

One of the most significant changes in the analysis of melting practice for aluminum scrap in the past 10 years is the use of computational fluid dynamics (CFD) to predict the impact of changes in process variables. Figure 7.5, from the work of Nieckele et al. (2011), demonstrates the value of CFD (Buchholz and Rødseth, 2011; Furu et al., 2010, 2012; Zhou et al., 2010). The two views show the wall temperature of a furnace being fired from the left with natural gas (Figure 7.5a) and a generic liquid fuel oil (Figure 7.5b). The higher wall temperatures near the outlet produced by firing with liquid fuel will shorten refractory life and result in uneven heating of aluminum in the furnace. CFD predicts a variety of things: the rate at which scrap placed in the furnace heats up, the temperature of the atmosphere in the furnace, heat loss through the furnace walls, the effect of changing furnace configurations and burner locations, the impact of using enriched air, and numerous other process variables (Matsuzaki et al., 2011; Murza et al., 2007). This promises to reduce development costs and allow rapid response to changing conditions (scrap and fuel prices, new burner designs).

CHEMISTRY OF FLUXING

All aluminum scrap fed to a furnace has a very thin skin of aluminum oxide on the surface, caused by the spontaneous oxidation of aluminum in air (Foseco, 2002):

$$2Al + \frac{3}{2}O_2 = Al_2O_3 \qquad (7.9)$$

The skin grows very slowly at normal temperatures and so is usually not apparent on the surface. When the scrap is heated in a furnace, the higher temperatures cause the skin to grow more rapidly, especially if the scrap is heated in air or an oxidizing environment. The addition of magnesium to the alloy causes the oxidation reaction to occur even faster and results in the production of spinel ($MgAl_2O_4$) as well as alumina (Eckert et al., 2005).

When the aluminum melts, the Al_2O_3 floats to the surface, forming a second phase known as *dross*. Because of the surface tension of the oxide skin, metallic aluminum is also trapped in the dross. The metal content of the dross can be anywhere from 40% to 80%, depending on melting conditions. While dross treatment processes recover most of this metal, at least some winds up being disposed of. As a result, reducing metal loss rates to the dross is an important consideration in scrap remelting (see Chapter 13).

One of the options available to remelters trying to reduce metal losses is to use a *drossing flux* designed to break the oxide skin and release the entrapped metal. Peterson (1990) lists the following requirements for such a flux:

- Should cover the molten metal and limit further oxidation
- Should dissolve or suspend dirt, oxides, and other nonmetallic substances
- Should promote coalescence of the metal droplets in the dross by stripping away the Al_2O_3 layer
- Should have a melting point lower than that of aluminum (660°C)
- Should have a density less than that of molten aluminum (2.3 g/cm³)
- Should not react with or contaminate molten aluminum
- Should not attack furnace refractories
- Should be relatively nontoxic
- Should have a low vapor pressure
- Should not be overly hygroscopic (avoid adsorbing water)
- Should be inexpensive and/or easily recycled
- Should have a low viscosity
- Should separate cleanly from the metal

These requirements quickly reduce the list of available materials to a short list of chloride and fluoride salts. Over time, a standard *cover flux* composition of 50 wt.-% NaCl, 50% KCl (43.9 mol-% NaCl) has become commonplace (Bolivar and Friedrich, 2009; Peterson, 1990). Figure 7.6 shows the NaCl–KCl phase diagram; a 50 wt.-% mixture has a liquidus temperature just above the eutectic temperature of 657°C. (An equimolar flux analyzing 56% KCl, 44% NaCl is also used.) However, while this flux is effective at covering the metal, it is less effective at causing coalescence. Furnace operators have learned that adding small levels of fluoride salt to the base cover flux promotes coalescence and generates a dross with much less metal in it. The most popular fluoride addition is *cryolite* (Na_3AlF_6), which is produced either synthetically or from used primary electrolysis cell bath (Bathco, 2012; Foseco, 2002). Sodium aluminum tetrafluoride (SATF, $NaAlF_4$) promotes coalescence even faster.

The reason why fluoride additions are effective remains a matter of some dispute. Initially, it was held that added fluoride increased the solubility of oxides in the drossing flux (Foseco, 2002), including Al_2O_3, and that the oxide skins simply dissolved. Most recent evidence, illustrated in Figure 7.7, suggests a different scenario (Besson et al., 2011; Friesen et al., 1997; Zholnin and Novichkov, 2005). In this view,

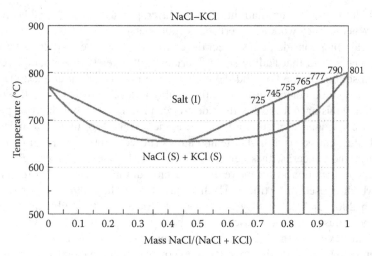

FIGURE 7.6 The NaCl–KCl phase diagram. (From Bolivar, L.R. and Friedrich, B., *World Metall.—Erzmetall.*, 62, 366, 2009. With permission.)

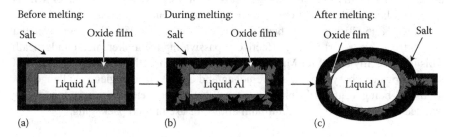

FIGURE 7.7 Impact of fluoride-containing fluxes on aluminum droplet coalescence in molten dross. (a) Metal adheres to the oxide film, (b) salt penetrates through the oxide layer, and (c) metal contracts and forms a spherical ball. (From Friesen, K.J. et al., Coalescence behaviour of aluminum droplets under a molten salt flux cover, in *Light Metals 1997*, Huglen, R., Ed., TMS-AIME, Warrendale, PA, p. 857, 1997. With permission.)

the changing shape of the aluminum in a droplet as it melts causes the oxide skin to crack. Molten flux penetrating the cracks makes contact with the molten droplet. It has been shown that fluoride additions (particularly Na_3AlF_6, NaF, and KF) greatly reduce the surface tension of molten flux on molten aluminum (Ho and Sahai, 1990), which causes the flux to spread out along the droplet surface. As this happens, the remnants of the oxide skin separate from the metal, freeing it to combine with other droplets as they come into contact. The exact mechanism may depend on the composition of alloy being melted.

A second way of encouraging coalescence is to heat the dross layer. A specific category of exothermic drossing fluxes has been developed that contains an oxidizing agent such as KNO_3, Na_2CO_3, and K_2SO_4, among others (Foseco, 2002). These react with the liquid aluminum in the dross layer, producing Al_2O_3 and heating the dross. Whether they work as advertised is controversial.

Several other considerations affect the composition of cover and drossing fluxes. It has been shown that adding some fluorides to flux tends to encourage reaction between sodium in the flux and the molten aluminum (Zholnin et al., 2005). This raises the sodium content of the metal, possibly to levels greater than allowed (see Chapter 12). Using AlF_3 as the fluoride source prevents this, but this is expensive. Another consequence of adding fluorides is the potential formation of undesirable fluoride dust emissions at higher temperatures (Foseco, 2002). In addition to work-place hygiene concerns, these emissions pose a risk to electric melting furnaces. A third consideration in recent years has been a dramatic increase in the price of KCl, which is used in fertilizers (Bolivar and Friedrich, 2009). Prices rose 250% between 2005 and 2008; they have since declined, but the vulnerability of the recycling industry to future price shocks has encouraged the development of low-potassium fluxes like the 70NaCl:30KCl mixture common in Europe.

Even with the best fluxing practice, some oxidation of the metal surface is bound to occur during the melting process. The dross limits heat transfer and thus slows melt rates. It also creates a risk of dross inclusions winding up in the molten metal and hurting product quality. As a result, preventing dross formation is an important characteristic of the melting process. Several dross prevention techniques will be described later in the book; one controversial technique is the addition of 0.0005–0.002 wt.-% (5–20 wppm) of beryllium (Eckert et al., 2005). The BeO layer that results from this is particularly effective at passivating the molten metal underneath. This is especially useful for Mg-containing alloys such as the 2000, 5000, and 7000 series. However, beryllium is a carcinogen and a workplace hygiene challenge, and its use is being voluntarily phased out by some producers. Its use is specifically avoided in the production of aluminum alloys for use in food packaging.

REFRACTORY INTERACTIONS

Although the demands on refractories in aluminum melting and holding furnaces are not as severe as those in other molten-metal processes, there are still some requirements:

- Lack of reactivity with the molten aluminum
- Low solubility in molten flux
- Lack of reactivity with the Al_2O_3 in the dross
- Low thermal conductivity
- Hydration resistance (important, since the atmosphere above the melt has a high water-vapor content from the combustion of the fuel)
- Oxidation resistance
- Stability at temperatures up to 1200°C (or higher if oxygen-enriched air is used in the burners)
- Good mechanical strength (important given the repeated loading of the fur-nace with solid scrap)

Of course, the most significant requirement is low cost.

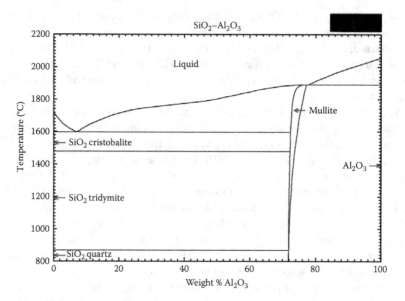

FIGURE 7.8 The alumina–silica phase diagram.

The combination of these requirements eliminates most of the available refractory compositions. Of the remaining choices, those which have by far become the most popular for melting and holding furnaces are refractories based on the alumina–silica system. Figure 7.8 illustrates the system, which features the ternary compound *mullite* ($Al_6Si_2O_{13}$) in addition to pure *corundum* (Al_2O_3) and *silica* (SiO_2). The refractories used in contact with molten metal and dross in aluminum melting and holding furnaces have a high alumina content (60%–85% Al_2O_3) and typically consist mostly of a mixture of corundum and mullite. However, there are variations in composition (CaO is often added to these refractories) and mineralogy (aluminosilicate refractories often contain significant levels of free silica). Recent developments in refractory composition and selection will be described in the next chapter.

The biggest concern in refractory use in molten aluminum is corrosion, especially at the melt line or *bellyband*. Corrosion results from the reaction between the molten aluminum and the refractory to generate corundum and elemental silicon (Guo et al., 2004; Yurkov and Pikhutin, 2009):

$$4Al + 3SiO_2 = 2Al_2O_3 + 3Si \tag{7.10}$$

$$4Al + \frac{3}{2}Al_6Si_2O_{13} = \frac{13}{2}Al_2O_3 + 3Si \tag{7.11}$$

Both reactions are thermodynamically favored at furnace temperatures. However, below the melt line the reaction is limited by the formation of a surface layer of corundum (Koshy et al., 2008), which prevents further access of the metal to the refractory underneath and stops the reaction. The surface layer is less adherent when the molten alloy contains higher levels of zinc or magnesium, and refractory corrosion is a greater concern in these conditions (Drouzy and Richard, 1974). Higher silica concentration in the refractory increases the reaction rate and thus the degree of corrosion.

At the bellyband, conditions are different, due to higher temperatures and the presence of oxygen in the adjacent furnace environment. As molten metal penetrates into the pores of the refractory, the oxygen reacts directly with the aluminum to form corundum (Allaire and Guermazi, 2000). The higher temperatures at the bellyband cause the reaction to occur even more quickly, and the porosity of the refractory allows the metal to migrate upward through the brick, forming corundum at locations above the melt line. The result is a corundum deposit large enough to reduce furnace capacity and adherent enough to take some of the original refractory with it when mechanically removed during cleaning (Raju, 2003). *Wall-cleaning fluxes* contain compounds such as sodium fluosilicate (Na_2SiF_6) and cryolite that react with the corundum and soften the deposit so that it can be scraped clean (Foseco, 2002).

A common practice in refractory development has been the addition of *nonwetting agents* to retard the interaction of the metal with the refractory (Koshy et al., 2008). Barite ($BaSO_4$), fluorspar (CaF_2), wollastonite ($CaSiO_3$), and aluminum fluoride (AlF_3) have all been used; boron-containing compounds also show promise (Adabifiroozjaei et al., 2012). The additives work by reacting with the refractory and the molten aluminum to generate multiple oxide phases such as anorthite ($CaAl_2Si_2O_8$), hexacelsian ($BaAl_2Si_2O_8$), and aluminum borate ($Al_{18}B_4O_{33}$), which reduce metal penetration into the refractory (Yurkov and Pikhutin, 2009); the reduced refractory loss justifies the cost of the additives. Wollastonite appears to be particularly helpful at reducing bellyband corrosion (Allaire and Guermazi, 2000).

RECOMMENDED READING

Besson, S., Friedrich, B., Pichat, A., Xolin, E., and Chartrand, P., Improving coalescence in Al-recycling by salt optimization, in *Proceedings of European Metallurgy Conference,* Vol. 3, GDMB, Clausthal–Zellerfeld, Germany, 2011, p. 759.

Lucas, C., Considerations in melting aluminum, Fives North American, September 2012. http://www.na-stordy.com/seminarCD/ppts-converted/AlumMelting.pdf.

Yap, L.T., Characteristics of flat–jet oxy–fuel flames for uniform heat transfer and low NOx emissions, in *Fourth Australasian-Asian Pacific Conference on Aluminium Cast House Technology*, Nilmani, M., Ed., TMS-AIME, Warrendale, PA, 1995, p. 67.

REFERENCES

Adabifiroozjaei, E., Koshy, P., and Sorrell, C.C., Effects of different boron compounds on the corrosion resistance of andalusite-based low-cement castables in contact with molten Al alloy, *Metall. Mater. Trans. B*, 43B, 5, 2012.

Alchalabi, R., Meng, F., and Peel, A., Furnace operation of optimization via enhanced bath circulation, in *Light Metals 2002*, Schneider, W., Ed., TMS-AIME, Warrendale, PA, 2002, p. 739.

Allaire, C. and Guermazi, M., Protecting refractories against corundum growth in aluminum treatment furnaces, in *Light Metals 2000*, Peterson, R.D., Ed., TMS-AIME, Warrendale, PA, 2000, p. 685.

Bathco, A Change of flux, March 7 2012. http://www.bathco.ch/files/bathco_brochure_e.pdf.

Besson, S., Friedrich, B., Pichat, A., Xolin, E., and Chartrand, P., Improving coalescence in Al-recycling by salt optimization, in *Proceedings of the European Metallurgical Conference 2005*, Vol. 3, GDMB, Clausthal–Zellerfeld, Lower Saxony, Germany, 2011, p. 759.

Bolivar L.R. and Friedrich, B., The influence of increased NaCl–KCl ratios on metal yield in salt bath smelting processes for aluminium recycling, *World Metall.—Erzmetall.*, 62, 366, 2009.

Buchholz, A. and Rødseth, J., Investigation of heat transfer conditions in a reverberatory melting furnace by numerical modeling, in *Light Metals 2011*, Lindsay, S.J., Ed., TMS-AIME, Warrendale, PA, 2011, p. 1179.

Chandler, R.C. and Shull, P.D., Increasing the surface emissivity of aluminum shapes to improve radiant transfer, in *Light Metals 2009*, Bearne, G., Ed., TMS-AIME, Warrendale, PA, 2009, p. 713.

Drouzy, M. and Richard, M., Oxydation des allages d'aluminium fondus, réaction avec les réfractaires, *Fonderie*, 332, 121, 1974.

Eckert, C.E., Meyer, T., Kinosz, M., Mutharasan, R., and Osborne, M., Preventative metal treatment through advanced melting technology, in *Shape Casting: The John Campbell Symposium*, Tityakioglu, M. and Crepeau, P.N., Eds., TMS-AIME, Warrendale, PA, 2005, p. 31.

Foseco, Development, evaluation, and application of granular and powder fluxes in transfer ladles, crucible and reverberatory furnaces, April 30 2002. http://www.foseco.com.tr/tr/downloads/FoundryPractice/237–02_Development_evaluation_appli_of_granular_flu-1.pdf.

Friesen, K.J., Utigard, T.A., Dupis, C., and Martin, J.P., Coalescence behaviour of aluminum droplets under a molten salt flux cover, in *Light Metals 1997*, Huglen, R., Ed., TMS-AIME, Warrendale, PA, 1997, p. 857.

Furu, J., Buchholz, A., Bergstrøm, T.H., and Marthinsen, K., Heating and melting of single Al ingots in an aluminum melting furnace, in *Light Metals 2010*, Johnson, J.A., Ed., TMS-AIME, Warrendale, PA, 2010, p. 679.

Furu, J., Buchholz, A., Bergstrøm, T.H., and Marthinsen, K., Numerical modeling of oxy-fuel and air-fuel burners for aluminium melting, in *Light Metals 2012*, Suarez, C.E., Ed., TMS-AIME, Warrendale, PA, 2012, p. 1037.

Gripenberg, H., Johansson, A., and Torvanger, K., Six years experience from low–temperature oxyfuel in primary and remelting aluminium cast houses, in *Light Metals 2012*, Suarez, C.E., Ed., TMS-AIME, Warrendale, PA, 2012, p. 1019.

Guo, J., Afshar, S., and Allaire, C., Corrosion kinetics of refractory by molten aluminum, in *Light Metals 2004*, Tabereaux, A.T., Ed., TMS-AIME, Warrendale, PA, 2004, p. 619.

Ho, F.K. and Sahai, Y., Interfacial phenomena in molten aluminum and salt systems, in *Second International Symposium on Recycling of Metal and Engineered Material*, van Linden, J.H.L., Stewart, D.L., and Sahai, Y., Eds., TMS-AIME, Warrendale, PA, 1990, p. 85.

Jenkins, R.F., Aluminum sidewall melting furnace heat transfer analysis, in *Fourth International Symposium on Recycling of Metal and Engineered Material*, Stewart, D.L., Stephens, R., and Daley, J.C., Eds., TMS-AIME, Warrendale, PA, 2000, p. 1045.

Koshy, P., Gupta, S., Sahajwalla, V., and Edwards, P., Effect of CaF_2 on interfacial phenomena of high alumina refractories with Al alloy, *Metall. Mater. Trans. B*, 39B, 603, 2008.

Li, T., Hassan, M., Kuwana, K., Saito, K., and King, P., Performance of secondary aluminum melting: Thermodynamic analysis and plant–site experiments, *Energy*, 31, 1769, 2006.

Lucas, C., Considerations in melting aluminum, Fives North American, September 2012. http://www.na-stordy.com/seminarCD/ppts-converted/AlumMelting.pdf.

Matsuzaki, K., Shimizu, T., Murakoshi, Y., and Takahashi, K., Evaluation of effects of stirring in a melting furnace for aluminum, in *Light Metals 2011*, Lindsay, S.J., Ed., TMS-AIME, Warrendale, PA, 2011, p. 1199.

Murza, S., Henning, B., and Jasper, H.-D., CFD-simulation of melting furnaces for secondary aluminum, *Heat Process.*, 5(2), 123, 2007.

Nieckele, A.O., Naccache, M.F., and Gomes M.S.B., Combustion performance of an aluminum melting furnace operating with natural gas and liquid fuel, *Appl. Therm. Eng.*, 31, 841, 2011.

Peterson, R.D., Effect of salt flux additives on aluminum droplet coalescence, in *Second International Symposium on Recycling of Metals and Engineered Materials*, van Linden, J.H.L., Stewart, D.L., and Sahai, Y., Eds., TMS-AIME, Warrendale, PA, 1990, p. 69.

Raić, K.T., Husović, T.V., and Jančić, R., Elements of refractory corrosion in secondary aluminium melting furnaces, *Metalurgija*, 10, 37, 2004.

Raju, R., Refractory selection, use and performance, in *Eighth Australasian Conference on Aluminium Cast House Technology*, Whiteley, P.R., Ed., TMS-AIME, Warrendale, PA, 2003, p. 47.

Solovjov, V.P. and Webb, B.W., Prediction of radiative transfer in an aluminium–recycling furnace, *J. Energy Inst.*, 78, 18, 2005.

Wechsler, T. and Gitman, G., Use of the Pyretron variable ratio air/oxygen/fuel burner system for aluminum melting, in *Energy Conservation Workshop XI: Energy and the Environment in the 1990s*, Aluminum Association, Washington, DC, 1990, p. 269.

Yap, L.T., Characteristics of flat–jet oxy–fuel flames for uniform heat transfer and low NOx emissions, in *Fourth Australasian-Asian Conference on Aluminium Cast House Technology*, Nilmani, M., Ed., TMS-AIME, Warrendale, PA, 1995, p. 67.

Yurkov, A.L. and Pikhutin, I.A., Corrosion of aluminosilicate refractories by molten aluminum and melts based upon it in melting and casting units, *Refr. Ind. Ceram.*, 50, 212, 2009.

Zholnin, A.B. and Novichkov, S.B., About the coalescence mechanism of aluminum, the analysis of the recent conceptions, in *Light Metals 2005*, Kvande, H., Ed., TMS-AIME, Warrendale, PA, 2005, p. 1197.

Zholnin, A.B., Novichkov, S.B., and Stroganov, A.V., Choice of additions to NaCl–KCl mixture for aluminum refining from alkali and alkaline–earth impurities, in *Light Metals 2005*, Kvande, H., Ed., TMS-AIME, Warrendale, PA, 2005, p. 973.

Zhou, N.-J., Zhou, S.-H., Zhang, J.-Q., and Pan, Q.-L., Numerical simulation of aluminum holding furnace with fluid–solid coupled heat transfer, *J. Cent. South Univ. Technol.*, 17, 1389, 2010.

8 Melting Furnace Parts and Accessories

As the next chapter will show, there are numerous designs of gas-fired melting furnaces for scrap aluminum. However, these furnaces have many parts in common. Advances in the design of these parts affect all types of furnace. These advances include the creation of *accessories*—equipment that is attached to the main body of the furnace, either for occasional or semicontinuous use. The purpose of these parts and accessories is to meet a common goal—melting aluminum scrap in the shortest possible time, with the highest possible efficiency and the lowest possible melt loss.

CHARGING AND PREHEATING

The traditional means of charging scrap to a reverb uses a front-end loader and an open furnace door (Magarotto, 2009; Williams, 2008). The method is simple, robust, and uncomplicated. However, it also exposes the furnace refractories to cold air (and thermal shock), causes damage to both the doors and the furnace as the front-end loader makes contact with it, and is a safety risk to the operator of the vehicle. The first step to improving this is replacing the front-end loader with a remote charging machine that removes the operator from the cab. Figure 8.1 shows such a vehicle, typically rail mounted (Emes, 2001). Several types are available; the choice of model depends on the type of scrap being charged. The machine can also be designed to extend all the way into the furnace so that scrap is dumped into the back as well as the front of the furnace cavity.

The next step forward is a vibratory feeder that acts as a buffer between the batch charging device and the furnace (Gillespie + Powers, Inc., 2003). These feeders are not inexpensive, but steady-state feeding and minimized thermal shock reduce refractory costs. Steady-state feeding also makes it possible to reduce the size of the doors, which reduces heat loss and improves workplace hygiene. The feed rate can potentially be controlled by the temperature of the melt (Grayson, 2011; Houghton et al., 2008), which minimizes temperature excursions and provides a slight improvement in productivity. Vibratory feeders are most often used for feeding small scrap such as auto shred or UBC.

Chapter 14 details the safety concerns associated with melting aluminum scrap, in particular explosions caused by wet or oily scrap. Some melting furnaces (see Chapter 9) make it possible to preheat scrap before melting, driving off the volatiles. For other melting furnaces (in particular wet-hearth and rotary furnaces), preheating scrap in a separate oven is a desirable option (Migchielsen and de Groot, 2006). These ovens are generally a simple arrangement, fired with natural gas (waste heat from the melting furnace can be used as a supplement). Depending on the type of scrap,

FIGURE 8.1   Rail-mounted charging device. (Courtesy Rob Nash, Gillespie + Powers, Inc.)

preheating temperatures as high as 500°C are used. If the scrap has a small level of oil or coatings, these preheating ovens can take the place of a decoating furnace.

BURNERS

The developers of improved burners try to meet several goals, some of which are at odds with each other:

- Improving thermal efficiency, which is lower than 30% for many furnaces
- Decreasing fuel costs
- Increasing the rate of heat transfer to the charge, which increases furnace productivity
- Decreasing melt loss
- Increasing refractory life, particularly at or above the melt line
- Decreasing emissions of pollutants, particularly NO_x, without generating excessive soot

Many new designs have been proposed. Some of the more important trends in burner design and operation are described in the following:

Use of oxygen: The difficulties caused by the 79% nitrogen in air were described in the previous chapter. The biggest of these difficulties is the heat lost in the nitrogen when it goes up the stack, which represents wasted energy and money (Kobayashi and Tsiava, 2003). High nitrogen content in the furnace atmosphere also increases the size of the ductwork needed to contain it and reduces flame temperatures, which

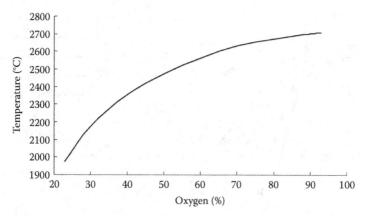

FIGURE 8.2   Impact of oxygen enrichment on natural-gas flame temperature. (Adapted from Becker, J.S. et al., in *EPD Congress 1997*, Mishra, B., Ed., TMS-AIME, Warrendale, PA, 1997. With permission.)

can be undesirable in some furnaces. As a result, some furnace operators now enrich their combustion air by adding oxygen. Figure 8.2 illustrates the impact of oxygen enrichment on flame temperature (D'Ettorre et al., 2007). As the previous chapter pointed out, hotter flames transfer heat more quickly to the metal underneath, improving productivity. The reduced nitrogen throughput of oxygen-enriched combustion means less heat lost in the nitrogen and thus higher efficiency; fuel savings of 30% or more can be achieved. Figure 8.3 illustrates another consequence of oxygen enrichment, higher flame velocity (Abernathy et al., 1996). As the previous chapter pointed out, heat transfer by convection is increased by higher gas turbulence. This

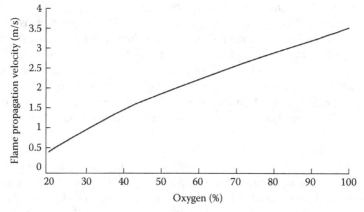

FIGURE 8.3   Impact of oxygen enrichment on natural-gas flame velocity. (Adapted from Abernathy, R. et al., The performance of current oxy–fuel technology for secondary aluminum melting, in *Light Metals 1996*, Hale, W., Ed., TMS-AIME, Warrendale, PA, 1996. With permission.)

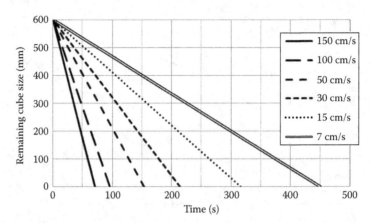

FIGURE 8.4 Melting speed of a 600 mm aluminum cube for different gas flow rates. (From Migchielsen and Gräb, 2007. With permission.)

is especially useful when heating solid aluminum. Figure 8.4 shows how much faster a 600 mm aluminum cube melts in hot gas flowing at a faster rate.

Potesser et al. (2008) describe four methods of using oxygen in burners for aluminum melting furnaces. The most common is *oxyfuel* burners that use pure O_2 and eliminate the use of air entirely. These nearly eliminate the presence of nitrogen in the system, minimizing off-gas volume and maximizing efficiency. However, the flame temperatures produced by these burners are too high, which increases dross formation and reduces refractory life (D'Ettorre et al., 2007). To solve this, some of the off-gas is recycled and mixed with the oxygen and natural gas. This lowers the flame temperature and spreads out the flame, providing heat transfer over a wider area. In recent years, this has given rise to *flameless combustion* (Gripenberg et al., 2007, 2012), in which the flame is so diffuse as to be largely invisible. The lack of N_2 in the off-gas increases the concentration of CO_2 and H_2O; these have higher emissivities than diatomic gases such as O_2 and N_2, and this increases the heat transfer by radiation. The lower temperature resulting from the wider distribution of the flame minimizes NO_x formation, and the higher fuel efficiency reduces CO_2 emissions, as seen in Figure 8.5 (Niehoff, 2009). Recycling off-gas also addresses another problem of using pure oxygen: the lower furnace pressure that results from the reduced amount of combustion product. This lower pressure can cause furnaces to leak, admitting nitrogen to the system and leading to NO_x generation (Jepson and van Kampen, 2005).

However, using pure oxygen costs money, while air is free. As a result, some burner designs have focused on ways to achieve the benefits of oxygen enrichment while using less of it. Burners that mix pure oxygen with air at varying levels of enrichment can achieve most of the benefits of oxyfuel burners while spending much less money on oxygen (D'Ettorre et al., 2007; Jepson and van Kampen, 2005; White, 2006). The input of nitrogen from the air means that off-gas volumes are larger, but the capital requirements for recirculation of flue gas are eliminated. NO_x emissions can be kept low if the O_2 content of the air is kept below 50% (Potesser et al., 2008).

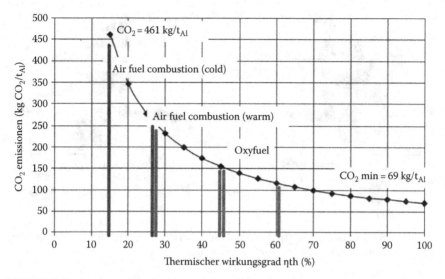

FIGURE 8.5 Impact of oxyfuel burners on CO_2 generation during reverberatory furnace melting. (From Niehoff, T., Optimised aluminium melting, presented at *10th OEA International Aluminium Recycling Congress*, Berlin, Germany, May 2, 2009, http://www.aluminium-recycling.com/de/verband/InvitationARC2009.pdf. With permission.)

To some degree, oxygen enrichment competes with regeneration (see below) as a device for reducing fuel requirements (White, 2006). Installation of oxyfuel or oxygen-enriched burners has a lower capital cost (Potesser et al., 2008), but the cost of oxygen makes this technology difficult to justify. As natural-gas prices decline, the economic return on oxyfuel burners or oxygen enrichment for aluminum melting furnaces declines as well.

Regeneration: Regeneration is a technique for heat recovery from the combustion off-gases, introduced to the secondary aluminum industry in the 1980s. Figure 8.6 illustrates the principle behind regenerative burners (Anon., 2009; Costa and Whipple, 2003). In this case, burners are mounted at opposite ends of the furnace. Each burner features a regeneration bed, comprised of a nonreactive ceramic (usually alumina balls, 2–3 cm diameter). When the burner at the left is in operation, exhaust gas is drawn through the regeneration bed at the right. The bed absorbs heat from the exhaust gas, heating up in the process. Every minute or so, the burner at the left is shut down and the one at the right fired up. When this happens, combustion air is drawn through the heated regeneration bed. The air is preheated by the bed, which loses a small amount of its energy in the process. At the same time, the regeneration bed at the left is opened to receive exhaust gas and is slightly heated up in turn. Another minute later, the left burner becomes operational again, and the process is repeated.

By preheating the air, regeneration increases flame temperatures, improving furnace productivity. Furthermore, the recovery of heat from the off-gas means that less heat is lost from the furnace; this decreases fuel costs by 40%–45% (Anon., 2009; Costa and Whipple, 2003). Because the impact of regeneration is similar to that of

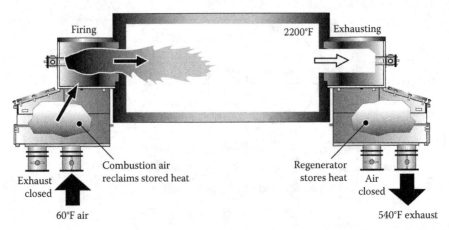

FIGURE 8.6 Use of regenerative burners in reverberatory melting furnaces. (From Fives North American Combustion, Inc. With permission.)

oxygen enrichment, the two techniques are rarely used at the same time. Regeneration also presents the same potential disadvantages as oxygen enrichment—higher melt loss, greater refractory wear, and high NO_x generation (Schalles, 1998). Similar to oxygen enrichment, recirculation of flue gases helps mitigate these concerns. An additional problem in furnaces that use flux is the tendency of the flux to form dust or vapor, which condenses in the regenerator. As a result, periodically the regenerators must be shut down for cleaning. This involves removing the ceramic media, discarding the contaminated material, and screening the rest to remove particulates and condensed dust. Furnaces using flux use larger alumina balls because of this.

Flame manipulation: Furnace operators have learned that different types of flame are desired at different points during the melting process. For example, during the early stages of melting, the low emissivity of metallic aluminum makes heat transfer by radiation difficult. A flame that transfers more energy by convection is desirable. After the metal is molten, the dross layer on top absorbs more radiation; in addition, the disturbance of the dross layer by turbulent exhaust gas increases melt loss. As a result, a flame with better radiation characteristics is preferred (Migchielsen and de Groot, 2005). In addition, a flame that radiates from an area rather than a single point will reduce the overheating of both metal and refractory.

In response to this, burner manufacturers have developed designs that manipulate flames in several ways (Abernathy et al., 1996; Costa and Whipple, 2003; Dentella and Zucca, 2011; Migchielsen and de Groot, 2005). A popular manipulation is the *flat flame*, in which the flame is spread out over some width, rather than an axisymmetrical cone. The *staged* introduction of air or fuel helps spread out the flame over an area, rather than a single location (Schalles, 1998). The staged introduction of air also encourages incomplete initial combustion of the fuel. This results in pyrolysis, as described in the previous chapter. The soot created by the pyrolysis makes the flame more luminous, encouraging heat transfer by radiation. Increasing the fuel and air velocity helps generate a *long* flame, which reduces overheating.

In addition to manipulating the size and shape of the flame, it is now increasingly possible to manipulate the direction of the flame (D'Ettorre et al., 2007; Migchielsen et al., 2010; Migchielsen and de Groot, 2005). This is useful in melting scrap, since the location of the piles of metal in a furnace can change from load to load. As the pile melts, changing the direction of the flame as well as the length becomes useful in reducing dross generation as well as producing even heating (Lucas, 2012). CFD analysis discussed in the previous chapter has become a useful tool for determining the optimal location for burners as well as how firing characteristics should change with time. Online analysis of the off-gas temperature and pressure is also commonly used to automatically control burner input and firing rate (Costa and Whipple, 2003; Hamers, 2007; Houghton et al., 2008). In some cases, the off-gas oxygen content is also analyzed for burner control (Migchielsen and de Groot, 2005).

RECUPERATORS

Recuperators perform a similar function to regenerators, in that they recover heat from the furnace exhaust gas and use it to preheat the combustion air. The main differences between the two are that (1) recuperators are a separate piece of equipment, rather than the integral part of the burner shown in Figure 8.6, and (2) recuperators operate on a steady-state basis, instead of the semicontinuous operation of regenerators.

Figure 8.7 illustrates a typical convection-type recuperator (Lucas, 2012). As the name suggests, the primary method of heat transfer in these devices is by convection from the hot exhaust to the surface of the metal tubes through which cold combustion air is input. Further, convection heat transfer from the inside surface of the tubes heats the combustion air. High-temperature Ni–Cr–Fe–Co alloys are used for tubes, largely to resist the corrosive nature of the off-gas (Gorog et al., 2007; Haynes International, 2008). However, the use of salt fluxes in melting furnaces creates problems in recuperators (Jepson and van Kampen, 2005), since halide vapors given off by the molten salt corrode even these alloys at high temperatures. As a result, the combustion air cannot be preheated above 350°C–400°C. This minimizes the possible fuel savings to about 20%. Recuperators have other disadvantages compared with regenerators. Their size per amount of heat recovered is large, and materials failure is a common problem. In addition, dust and condensing salt vapor frequently clog the tubes, requiring shutdowns. As a result, they are rarely used in new furnace installations, although they have some value in small batch melting furnaces (Nabertherm, 2012).

REFRACTORIES

The challenges to refractory performance in aluminum melting and holding furnaces were described in the previous chapter. Responses vary from operator to operator, but some trends are apparent:

- A shift away from bricks toward monolithic linings (Cölle and Schiffbauer, 2012; Houghton et al., 2008; Whiteley, 2011). These are cheaper to install and have no joints to penetrate. However, monolithic linings do not have

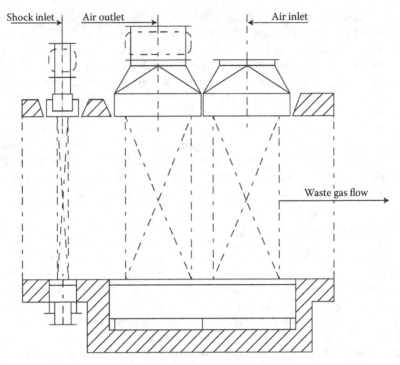

FIGURE 8.7 Schematic of recuperator with shock-tube bundles. (From Ottie, T.W., in *Energy Conservation Workshop XI: Energy and the Environment in the 1990s,* Aluminum Association, Washington, 1990. With permission.)

the same mechanical strength as bricks and so are more easily damaged when the charge is dumped into the furnace. Figure 8.8, the internal structure of a generic stationary melting furnace (Wynn et al., 2011b), shows the areas of refractory use. The hearth and ramp are the most frequent cause of shutdowns, due to mechanical damage, abrasion, and thermal shock during loading of cold scrap. As a result, bricks are still used in large furnaces subjected to heavy loading (Decker, 2011), such as round furnaces (see next chapter). In areas above the melt line, alkali attack from flux dust and vapor is a cause for concern (Wynn et al., 2011a). The development of low-cement castables and gunning mix has been especially significant (Carden and Brewster, 2008). Over the past 10 years, the use of precast sections (*big blocks*) has become more popular (Jones and Taberham, 2008; Pyrotek, 2005). These cost more and require longer lead times than castable linings, but precasting allows for more consistent curing, reduces downtime, and increases the range of available refractory compositions.

• The use of nonwetting agents in linings to limit metal penetration (Cölle and Schiffbauer, 2012). Barium sulfate is the most common of these; calcium fluoride is often used as well (Hemrick et al., 2008). The effectiveness of

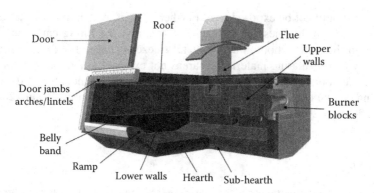

FIGURE 8.8 Internal structure of a stationary melting furnace. (From Wynn, A. et al., Improved monolithic materials for lining aluminum holding and melting furnaces, in *Light Metals 2011*, Lindsay, S.J., Ed., TMS-AIME, Warrendale, PA, p. 663, 2011b. With permission.)

these agents tends to decrease at melt temperatures above 1100°C (Bonadia et al., 2006). Silicon carbide is also a useful antiwetting addition, but increases the thermal conductivity of the refractory.

- Higher quality refractory. Eliminating silica fume is especially important for reducing corundum formation (Hemrick et al., 2008). However, this is difficult, since the silica is needed to ensure flowability in castable mixes. Efforts to replace the silica with kyanite are continuing.
- Composite refractory walls such as that in Figure 8.9 (Morrow, 2003), with a corrosion-resistant refractory at the hot face and a low-density insulating material underneath to reduce heat losses. It is important to make sure that the high-temperature refractory is thick enough that the freeze plane below which penetrating metal solidifies remains within it (Migchielsen et al., 2010). Insulating brick is less resistant to molten metal

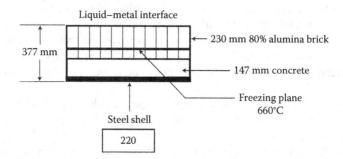

FIGURE 8.9 Composite monolithic refractory structure for reverberatory melting furnace. (From Morrow, R., Development of refractory linings in holding furnaces at Tomago Aluminium, in *Aluminium Cast House Technology—Eighth Australasian Conference*, Whiteley, P.R., Ed., TMS-AIME, Warrendale, PA, p. 101, 2003. With permission.)

and should not be exposed to it. Furthermore, if molten metal penetrates beyond the thickness of the facing refractory, the lining can *heave up* from the wall or bottom as the metal freezes between the facing refractory and insulating material. In areas not exposed to molten metal such as the doors, roof, and upper walls, this is less of a concern, and the lining can be chosen with insulating qualities more in mind (White, 2006; Whiteley, 2011).

STIRRING

The term *stirring* encompasses a wide range of technologies, often designed for specific furnace architectures. Stirring is meant to accomplish one or more of several goals (Bright et al., 2007a; Campbell, 2010; Guest et al., 2012; Henderson et al., 1996):

- *Elimination of thermal gradients*: Because heat is transferred to the charge in a melting furnace entirely from the top, the temperature of the melt at the top can be up to 100°C higher than that at the bottom. Thermal gradients lead to excessive dross formation and melt loss (the metal is too hot at the top), variable chemistry, and problems with casting when metal of varying temperatures is tapped. Stirring evens out the metal temperature, improving product quality and reducing melt loss. It also reduces the temperature at the metal surface, which in turn lowers refractory temperatures above the melt and reduces refractory wear.
- *Submergence of light scrap*: Light scrap such as UBCs or foil tends to float on top of molten metal when charged. Leaving it there leads to excessive melt loss, so submerging it under the melt as quickly as possible is important.
- *Elimination of inclusions*: While good quality control in cast shops now requires filtration, the larger inclusions in molten aluminum can be eliminated beforehand either by flotation on bubbles of introduced argon or by attachment to the furnace walls. Stirring makes this easier to accomplish.
- *Higher melt rate*: Stirring increases the convective heat transfer coefficient h, improving heat transfer from the top of the melt to the metal or scrap below. This increases melt rates by 20% or more compared to unstirred furnaces.

The numerous devices used for stirring molten metal can be separated into four basic types of device: *bubble injection, mechanical pumps, electromagnetic pumps*, and *electromagnetic stirrers*.

Bubble injection is accomplished in one of two ways: through a porous plug in the furnace bottom (Migchielsen et al., 2010) or through a lance or fluxing wand inserted through the top. The bubbles are argon or an argon–chlorine mixture and are injected primarily for degassing or fluxing rather for stirring. Recently introduced rotary flux injectors are more effective for stirring but again are primarily meant for other purposes (see Chapter 12).

FIGURE 8.10 Photograph of improved mechanical pump for molten aluminum stirring. (From Tensor™ Series Pumps Website, http://www.metaullics.com/tensor.html. With permission.)

Most of the mechanical pumps in use for molten aluminum service are centrifugal (see Figure 8.10) rather than positive displacement, due to the higher discharge volumes and velocities of centrifugal designs. These pumps cost less per unit capacity than other stirring devices and cost less to operate as well (Henderson et al., 1996); maintenance costs are higher, but development work has focused on improving their reliability (Bright et al., 2007b; Chandler, 2006). Mechanical pumps are often portable and can be used on more than one furnace. The newest pumps can circulate 20,000 kg/min of molten aluminum, with less turbulence than previous lower-capacity designs. Centrifugal pumps are usually used only on open-well furnaces, however; closed-system furnaces require a different type of stirrer. They also have difficulty in furnaces where metal drains away from the pump as the furnace is emptied and so are less desirable for round and dry-hearth furnaces. Nevertheless, their low cost and simplicity of operation make them a competitive option.

Figure 8.11 illustrates the operating principle behind the electromagnetic pump, introduced to the aluminum industry about 15 years ago (EMP Technologies, 2005; Henderson et al., 1996; Peel, 2003). The equipment consists of a well, lined with the same high-alumina refractories used for the furnace itself, and the pump itself, wrapped around a silicon carbide inlet tube. The pump consists of a coil wrapped around the tube, through which mains-frequency alternating current flows; the forces exerted on the metal in the tube are the same as those described for the eddy

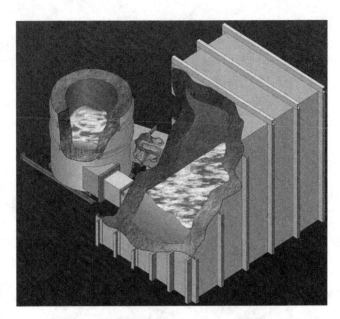

FIGURE 8.11 Illustration of electromagnetic pump attached to melting furnace. (From EMP System > How it Works website, http://www.emptechnologies.com/emp/dom93/index. htm. With permission.)

current separator in Chapter 6. In this case, exerting this type of force on molten metal causes it to flow through the tube. (The same principle is behind the operation of induction furnaces, described in Chapter 10.) Electromagnetic pumps are more expensive to install but are more effective than mechanical pumps in stationary melters, reduce operating costs, and are much easier to maintain (EMP Technologies, 2007; Pyrotek, 2005). As a result, they are the primary choice for new installations.

Electromagnetic stirrers use a similar principle to the induction furnaces used for melting (see Chapter 10). Mounted externally to the furnace, they consist of an iron core adjacent to a set of water-cooled stirrer coils (Peel, 2003). AC electric current passed through the stirrer coils generates a magnetic field, which in turn generates a secondary electric current in the molten metal in the furnace. As the current oscillates, the changing magnetic field in the metal generates eddy currents, which stir the metal. As with induction furnaces, low frequencies generate better depth of penetration of the magnetic field, allowing use of refractories of normal thickness. Electromagnetic stirrers can be mounted either under the furnace, as shown in Figure 8.12, or to the side. Side-mounted stirrers are the typical choice for retrofitted furnaces. To allow the magnetic field to penetrate into the furnace, the carbon steel shell of the furnace is replaced by a window of nonmagnetic austenitic stainless steel where the stirrer is located (Stål and Hanley, 2009). A bath height of at least 0.33 m is recommended.

Electromagnetic stirrers are more effective than any other stirring device, which maximizes the gains in productivity and product quality that can be achieved. As a result, their use has spread rapidly over the past decade, as shown in Figure 8.13

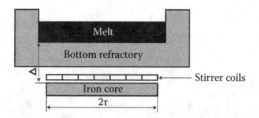

FIGURE 8.12 Schematic of bottom-mounted electromagnetic stirrer. (From Eidem, M. et al., Side mounted EMS for aluminum scrap melters, in *Light Metals 1996*, Hale, W., Ed., TMS-AIME, Warrendale, PA, 1996. With permission.)

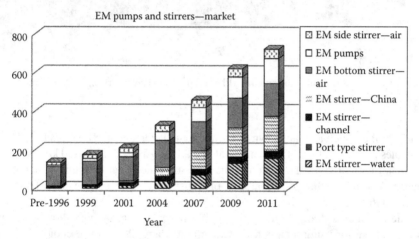

FIGURE 8.13 Increasing use of different types of electromagnetic pumps and stirrer. (After Peel, A. and Herbert, J., *Technology for Electromagnetic Stirring of Aluminum Reverberatory Furnaces*, Derbyshire, U.K., December 2, 2010, http://www.altek-al.com/downloads/Technology-for-Electromagnetic-Stirring-of-Aluminium-Reverberatory-Furnaces.pdf.)

(Peel and Herbert, 2010). Recent innovations include the use of solid copper inductors, rather than hollow copper coil; this eliminates the need for water cooling. The use of side-mounted stirrers as an alternative for pumps in vortex and sidewell furnaces has also been demonstrated (Guest et al., 2012). These operate at lower frequencies (5–10 Hz), producing better magnetic field depth and producing better results with less power use. Questions about worker-safety impacts of the electromagnetic field generated by the stirrer are a concern, as is the relatively high cost of these devices. However, their advantages promise to make them an increasingly common part of the melting-furnace landscape.

ROAD CRUCIBLES

As Chapter 11 points out, it is common for aluminum recycling operations to be located near their customers. Because of this, some recyclers have begun the practice of shipping aluminum to their customers in molten form, using specially designed

FIGURE 8.14   Fifteen-ton road crucible. (Photo courtesy of Ray Peterson, Aleris International.)

road crucibles such as that shown in Figure 8.14 (Peterson and Blagg, 2000). These crucibles hold 8–10 tons of molten metal and are transported up to 300 km. Keeping the metal molten saves the customer the cost and time of remelting it, which more than makes up for the additional cost of transporting it in molten form (Bonadia et al., 2006).

The crucible is drained either by pouring from the top, tapping through a plug in the bottom, or using a siphon (Peterson and Blagg, 2000). The choice depends largely on the customer's available equipment. As the distances over which molten metal is shipped increase, means of heating the metal to keep warm increasingly go along for the ride. Both electric and gas-fired portable heating units are available; the gas-fired units seem to be more economical (Prillhofer and Knaack, 2012). When crucibles arrive at their destination, temperature stratification has often set in; bottom-mounted electromagnetic stirrers can be used for these vessels as well as stationary furnaces (Stål and Hanley, 2009).

ENVIRONMENTAL EQUIPMENT

Reverberatory furnaces can emit a variety of interesting substances when melting aluminum scrap. The most significant of these are chloride gases and vapors. Scrap that has not been decoated often contains chlorinated polymer coatings, such as PVC. Furthermore, the salt used for fluxing is a KCl–NaCl mixture and gives off vapors of its own when molten. Improperly combusted fuel is a potential source of volatile organic compounds (VOCs) and may react with chloride vapors to generate chlorinated organic vapor compounds such as furans or dioxins (see Chapter 14). Along with this, scrap charging often generates dust that must be dealt with.

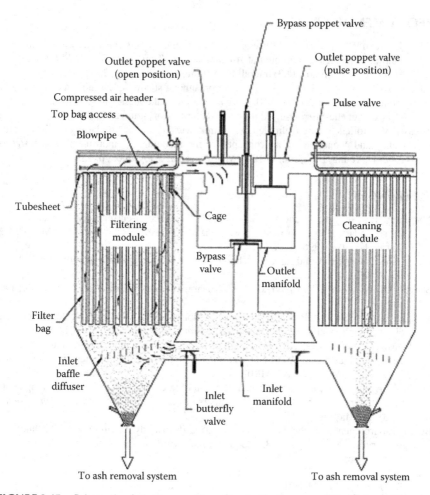

FIGURE 8.15 Schematic of a baghouse. (From Bundy Environmental Technology, *Filtering Module*, Bundy Environmental Technology, Westerville, OH, April 28, 2008, http://www. bundyenvironmental.com/syscomp.html#baghouse. With permission.)

Figure 8.15 shows a typical baghouse used to filter the cooled off-gas (Bundy Environmental Technology, 2008; CECO Environmental, 2010; Merz, 2004). The bags are large fabric filters, capable of removing micron-size dust particles from furnace exhaust gas. Polyesters are typically used for the fabric in aluminum remelting plants, due to their resistance to the HCl in the exhaust gas. However, their relatively low limiting operation temperature requires that gases be diluted with air to cool them below this temperature before being filtered. The bags are often coated with lime, to help recover chlorides in the off-gas. Nomex or other aramid fibers are used as the fabric for higher-temperature gases; these cost more than polyester, but the lower volume of off-gas that requires filtering when diluted air is not used reduces the required bag size and thus the cost.

RECOMMENDED READING

Costa, D.S. and Whipple, D.F., The use of regenerative burners systems to improve aluminium melting efficiencies, presented at *Aluminium Two Thousand: 5th World Congress on Aluminum*, March 20, 2003, Interall Publishing, Modena, Italy, 2003.

Peel, A. and Herbert, J., Technology for electromagnetic stirring of aluminum reverberatory furnaces, December 2, 2010. http://www.altek-al.com/downloads/Technology-for-Electromagnetic-Stirring-of-Aluminium-Reverberatory-Furnaces.pdf.

Potesser, M., Holleis, B., Hengelmolen, A., Antrekowitsch, H., and Spoljaric, D., Highest efficiency and economic burner concepts for the metallurgical industry, in *Light Metals 2008*, DeYoung, D.H., Ed., TMS-AIME, Warrendale, PA, 2008, p. 553.

REFERENCES

Abernathy, R., McElroy, J., and Yap, L.T., The performance of current oxy–fuel technology for secondary aluminum melting, in *Light Metals 1996*, Hale, W., Ed., TMS-AIME, Warrendale, PA, 1996, p. 1233.

Anon., Energy efficiency and furnace technology in modern foundries, *Aluminium*, 82(6), 46, June 2009.

Becker, J.S. et al., in *EPD Congress 1997*, Mishra, B., Ed., TMS-AIME, Warrendale, PA, 1997.

Bonadia, P., Braulio, M.A.P., Gallo, J.B., and Pandolfelli, V.C., Refractory selection for long distance molten–aluminum delivery, *Am. Ceram. Soc. Bull.*, 85(8), 9301, 2006.

Bright, M.A., Chandler, R.C., and Henderson, R.S., Advances in molten metal pump technology expand the capability of aluminum reverberatory production rates, in *Light Metals 2007*, Sørlie, M., Ed., TMS-AIME, Warrendale, PA, 2007a, p. 603.

Bright, M.A., Chandler, R.C., Henderson, R.S., and Starczewski, R.J., Recent developments in molten metal pumps to enhance reverberatory furnace productivity, *Alum. Int. Today*, 19(3), 19, May/June 2007b.

Bundy Environmental Technology, *Filtering Module*, Bundy Environmental Technology, Westerville, OH, April 28, 2008. http://www.bundyenvironmental.com/syscomp.html#baghouse.

Campbell, P., The benefits of forced circulation for aluminium reverberatory furnaces, *Mater. Sci. Forum*, 630, 111, 2010.

Carden, Z. and Brewster, A., Monolithic refractory linings designed for rapid commissioning, in *Light Metals 2008*, DeYoung, D.H., Ed., TMS-AIME, Warrendale, PA, 2008, p. 587.

CECO Environmental, *Aluminum Recycling Plant Turns to CECO Environmental for New Furnace Dust Collection System*, CECO Environmental, Cincinnati, OH, November 8, 2010. http://www.cecoenviro.com/filters/CaseHistories/CleaningTheAir.html.

Chandler, R.C., Current developments in molten metal pumps and their application in aluminium scrap recycling, presented at *3rd International Conference on Aluminium Recycling*, Moscow, Russia, March 30, 2006. http://www.pyrotek.info/documents/techpapers/MoscowPaper2006-final.pdf.

Cölle, D. and Schiffbauer, G., State-of-the-art monolithics for the aluminium smelting industry—Economic viability and reliability, *Refract. Worldforum*, 4(3), 10, March 2012.

Costa, D.S. and Whipple, D.F., The use of regenerative burners systems to improve aluminium melting efficiencies, presented at *Aluminium Two Thousand: 5th World Congress on Aluminum*, March 20, 2003, Interall Publishing, Modena, Italy, 2003.

D'Ettorre, A., Giudici, R., Tasca, A., Visus, J., Benavides, V.G., and Kobayashi, W.T., Low dross generation with oxy-fuel system, *Metall. Ital.*, 99(6), 47, 2007.

Decker, J., Phosphate bonded monolithic refractory materials with improved hot strength as a potential replacement for phosphate bonded bricks, *Mater. Sci. Forum*, 693, 90, 2011.

Dentella, F. and Zucca, P., Delivering flexible furnace operation with oxygen and oxy-fuel combustion techniques, presented at *Aluminium Two Thousand: 7th World Congress on Aluminum*, May 17, 2011, Interall Publishing, Modena, Italy, 2011.

Eidem, M., Tallbäck, G., and Hanley, P.J., Side mounted EMS for aluminum scrap melters, in *Light Metals 1996*, Hale, W., Ed., TMS-AIME, Warrendale, PA, 1996.

Emes, C.B., Improvements in metal quality and operating efficiency through furnace design, presented at *International Melt Quality Workshop*, Madrid, Spain, October 25, 2001.   http://www.pyrotek.info/documents/techpapers/Metal-Quality-thru-Furnace-Design-CBEmes.pdf.

EMP System > How it Works website. http://www.emptechnologies.com/emp/dom93/index.htm.

EMP Technologies, *EMP System Overview*, EMP Technologies, Staffordshire, U.K., May 10, 2005. http://nstg.nevada.edu/mmrg/research/LitSurvey/PDFs/EMP-119.pdf.

EMP Technologies, Raising the level of electromagnetic pumping, *Alum. Int. Today*, 19(3), 24, May/June 2007.

Fives North American Combustion, Inc.

Gillespie + Powers, Inc., *Vibratory Scrap Feeder MHS*, Gillespie + Powers, Inc., Saint Louis, MO, April 2, 2003. http://www.gillespiepowers.com/brochures/VBINtechwriteup.pdf.

Gorog, J.P. et al., *Final Technical Report: Materials for Industrial Heat Recovery Systems. Task 1: Improved Materials and Operation of Recuperators for Aluminum Melting Furnaces*, September 30, 2007. http://www.ornl.gov/sci/ees/itp/documents/FnlRptRecuperatorsFinal.pdf.

Grayson, J.T., Cost savings in the case house through optimizing furnace operation, staff training and associated variables, *Mater. Sci. Forum*, 693, 104, 2011.

Gripenberg, H., Johansson, A., Eichler, R., and Rangmark, L., Optimised oxyfuel melting process at SAPA Heat Transfer AB, in *Light Metals 2007*, Sørlie, M., Ed., TMS-AIME, Warrendale, PA, 2007, p. 597.

Gripenberg, H., Johansson, A., and Torvanger, K., Six years experience from low–temperature oxyfuel in primary and remelting aluminium cast houses, in *Light Metals 2012*, Suarez, C.E., Ed., TMS-AIME, Warrendale, PA, 2012, p. 1019.

Guest, G., Williams S., and Gastaldi, P., Development of a new generation electromagnetic metal moving system, in *Light Metals 2012*, Suarez, C.E., Ed., TMS-AIME, Warrendale, PA, 2012, p. 1013.

Hamers, C., Modernisation of melting and casting furnaces, *Aluminium*, 80(6), 60, 2007.

Haynes International, *Haynes® 556® Alloy Information*, Haynes International, Kokomo, IN, October 23, 2008. http://www.haynesintl.com/Haynes556Alloy/556HaynesAlloySR.htm.

Hemrick, J.G., Headrick, W.L., and Peters, K.-M., Development and application of refractory materials for molten aluminum applications, *Int. J. Appl. Ceram. Technol.*, 5, 265, 2008.

Henderson, R.S., Chandler, R.C., and Brown, W., A new electromagnetic circulation pump for aluminum reverberatory furnaces, in *Light Metals 1996*, Hale, W., Ed., TMS-AIME, Warrendale, PA, 1996, p. 869.

Houghton, B., Augustine, S., and Guest, G., Melting furnaces, technical developments reviewed and thoughts for the future, in *Light Metals 2008*, DeYoung, D.H., Ed., TMS-AIME, Warrendale, PA, 2008, p. 541.

Jepson, S. and van Kampen, P., Oxygen–enhanced combustion provides advantages in Al-melting furnaces, *Ind. Heat.*, 72(6), 29, June 2005.

Jones, D. and Taberham, S., The use of large precast refractory shapes for lining aluminium melting and holding furnaces, presented at *4th International Melt Quality Workshop*, Istanbul, Turkey, May 21, 2008. http://www.improvingperformance.cpm/papers/2_TAB.pdf.

Kobayashi, H. and Tsiava, R., Oxy–fuel burners, in *Industrial Burners Handbook*, Baukal, C.E., Jr., Ed., CRC Press, Boca Raton, FL, 2003, p. 693.

Lucas, C., *Considerations in Melting Aluminum,* Fives North American, Cleveland, OH, September 2012. http://www.na-stordy.com/seminarCD/ppts-converted/AlumMelting. pdf.

Magarotto, G., Melting shop loaders offer greater efficiency, *Alum. Int. Today*, 21(3), 26, May/ June 2009.

Merz, S.K., *Dioxin and Furan Emissions to Air from Secondary Metallurgical Processes in New Zealand*, April 2004. http://www.mfe.govt.nz/publications/hazardous/dioxin-furan-emissions-vol-1/dioxin-furan-emissions-vol1-apr04.pdf.

Migchielsen, J. and de Groot, J., Potentials for increasing fuel efficiency for aluminium melting furnaces, in *Light Metals 2005*, Kvande, H., Ed., TMS-AIME, Warrendale, PA, 2005, p. 893.

Migchielsen, J. and de Groot, J., Design considerations for charge preheating ovens, in *Light Metals 2006*, Galloway, T.J., Ed., TMS-AIME, Warrendale, PA, 2006, p. 737.

Migchielsen, J., Gräb, H.-W. and Schmidt, T., Retrofitting aluminum melting furnaces, in *Light Metals 2010*, Johnson, J.A., Ed., TMS-AIME, Warrendale, PA, 2010, p. 663.

Migchielsen, J. and Gräb, H.-W. 2007.

Morrow, R., Development of refractory linings in holding furnaces at Tomago Aluminium, in *Aluminium Cast House Technology—Eighth Australasian Conference,* Whiteley, P.R., Ed., TMS-AIME, Warrendale, PA, 2003, p. 101.

Nabertherm, *Foundry,* June 13, 2012. http://www.nabertherm.com/produkte/giesserei/ giesserei_english.pdf.

Niehoff, T., Optimised aluminium melting, presented at *10th OEA International Aluminium Recycling Congress,* Berlin, Germany, May 2, 2009. http://www.aluminium-recycling.com/ de/verband/InvitationARC2009.pdf.

Ottie, T.W., in *Energy Conservation Workshop XI: Energy and the Environment in the 1990s,* Aluminum Association, Washington, DC, 1990.

Peel, A.M., A look at the history and some recent developments in the use of electromagnetic devices for improving operational efficiency in the aluminium cast house, in *Aluminium Cast House Technology—Eighth Australasian Conference*, Whiteley, P.R., Ed., TMS-AIME, Warrendale, PA, 2003, p. 71.

Peel, A. and Herbert, J., *Technology for Electromagnetic Stirring of Aluminum Reverberatory Furnaces*, Derbyshire, U.K., December 2, 2010. http://www.altek-al.com/downloads/ Technology-for-Electromagnetic-Stirring-of-Aluminium-Reverberatory-Furnaces.pdf.

Peterson, R.D. and Blagg, G.G. Transportation of molten aluminum, in *4th International Symposium on Recycling of Metals and Engineered Materials,* Stewart, D.L., Stephens, R., and Daley, J.C., Eds., TMS-AIME, Warrendale, PA, 2000, p. 857.

Potesser, M., Holleis, B., Hengelmolen, A. Antrekowitsch, H., and Spoljaric, D., Highest efficiency and economic burner concepts for the metallurgical industry, in *Light Metals 2008*, DeYoung, D.H., Ed., TMS-AIME, Warrendale, PA, 2008, p. 553.

Prillhofer, B. and Knaack, J., Influence of heating technology on melt quality in ladles for road transportation of liquid aluminum casting alloys during holding, in *Light Metals 2012*, Johnson, J.A. Ed., TMS-AIME, Warrendale, PA, 2012, p. 633.

Pyrotek, *Improving Performance in Casthouse Furnaces,* August 29, 2005. http://www. pyrotek.info/furnace_operations.

Schalles, D.G., The next generation of combustion technology for aluminum melting, in *Light Metals 1998,* Welch, B., Ed., TMS-AIME, Warrendale, PA, 1998, p. 1143.

Stål, R. and Hanley, P., Electromagnetic stirring in aluminium ladles, in *Light Metals 2009*, Bearne, G., Ed., TMS-AIME, Warrendale, PA, 2009, p 627.

Tensor™ Series Pumps. Website. http://www.metaullics.com/tensor.html.

White, D.W., Furnace energy saving myths, The Schaefer Group, Melbourne, Victoria, Australia, April 13, 2006. http://www.fwschaefer.com/document/WhitePaper2006-300dpi.pdf.

Whiteley, P., A historical perspective of aluminium casthouse furnace developments, *Mater. Sci. Forum,* 693, 79, 2011.

Williams, K., Increased melt furnace productivity and other benefits of dedicated rail–bound furnace charging machines, *Aluminium,* 81(3), 48, March 2008.

Wynn, A., Coppack, J., Steele, T., and Latter, G., Improved monolithic materials for lining aluminium holding and melting furnaces—Roof, upper walls and flue, *Mater. Sci. Forum,* 693, 80, 2011a.

Wynn, A., Coppack, J., Steele, T., and Moody, K., Improved monolithic materials for lining aluminum holding and melting furnaces, in *Light Metals 2011,* Lindsay, S.J., Ed., TMS-AIME, Warrendale, PA, 2011b, p. 663.

9 Fossil-Fuel Furnaces

The reverberatory furnace used to melt aluminum scrap has numerous designs and sizes, and new designs appear on a regular basis. However, new furnaces are expensive, and so many older designs are still widely used. The goal of furnace design is consistent, however: to generate the highest melting capacity per unit volume, while maximizing thermal efficiency to reduce fuel costs. Other considerations include the ability of the furnace to handle finely divided scrap without excessive melt loss, the ability to process coated scrap if prior decoating is unavailable, minimizing maintenance costs, and ease of charging and tapping.

There are a number of ways to classify furnace designs. This chapter will separate them into *single-chamber* and *multiple-chamber* types. Additional sections at the back will describe two specialized types of gas-fired furnace: the *rotary* and *crucible* furnace. For more complete description of furnace designs, the reader is encouraged to read the review by Schmitz (2006).

SINGLE-CHAMBER DESIGNS
(KEARNEY, 1988; SCHAEFER FURNACES, 2008)

The *wet-hearth* furnace described in the previous edition of this book has become largely obsolete, for reasons described by Newman (2010). As a result, the most common type of single-chamber furnace is the *dry-hearth* furnace shown in Figure 9.1. This furnace features a sloping hearth on the right onto which solid scrap is placed for initial heating. As the metal melts, it drains down the hearth into the bath on the left, leaving other metallic materials behind (Boeckenhauer and Kaczmarczyk, 2008; Kennedy, 2001). The scrap also dries during the heating process, reducing the possibility of explosions and also reducing the potential for melt loss from the interaction between the metal and water vapor. When the molten metal reaches the desired temperature, it is either pumped out or tapped from a hole in the side. Burners are mounted in the opposite end of the furnace from the charge door; mounting the flue near the charge door gives hot combustion gases more chance to transfer energy to the melt as they head toward the flue. Burners can be roof or side mounted; for larger-capacity furnaces, regenerative burners or the use of preheated air is advantageous (Valder, 2011). Without the use of stirring or other melting aids, melt rates of 200 kg/h-m² are typical.

Metal removal from dry-hearth and other furnaces is accomplished one of three ways. The use of a hydraulic pump is the oldest (Brooks, 1970) and is used mostly for furnaces that are not integrated into a melting/holding/casting line. However, the pumps require maintenance and periodic replacement, and safety is a concern. Electromagnetic pumps were described in the previous chapter as a means of stirring the melt; these can also be used to transfer molten metal from the furnace and are increasingly used for that purpose (EMP Technologies, 2007; Starczewski, 2012).

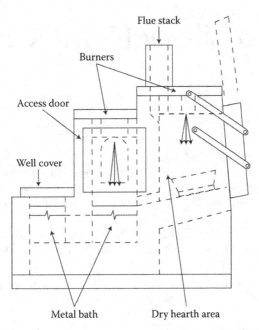

FIGURE 9.1   Dry-hearth melting furnace. (From McKenna, J.P. and Wisdom, A., *Die Cast. Eng.*, 41(6), 64, June 1997. With permission.)

These pumps require a certain level of molten metal in the furnace to operate and so cannot be used to completely drain the vessel.

A more common tapping approach for reverbs requires a hole drilled through the furnace side and filled with a quick-setting mortar plug or tap cone. When the molten metal is ready to tap, the plug is removed and replaced by a tap that allows metal to drain into a launder or transfer ladle. When the furnace is mostly empty, a new plug or cone is mortared into place. Tapping can be slow, especially toward the end of the tap when molten metal levels in the furnace are low. Furthermore, the plug material can be a source of inclusions in the metal.

The newest tapping approach involves tilting the entire furnace up, causing metal to drain through a pouring spout located at the front. Tilting has a significant power requirement, and costs more than tapping arrangements. As a result, it is rarely used for large melting furnaces. At the same time, it offers greater control of drainage rate than tapping and reduced inclusion content in the metal. Because of this, furnaces that tip are increasingly chosen for new installations, especially for rotary and holding furnaces (Barry et al., 2009; Emes, 2001).

An early improvement in furnace design was the development of designs with wide charging doors, allowing much larger loads to be placed in the furnace than before. However, charging furnaces this way is slow. The next step was the design of a *round* melting furnace (Boeckenhauer and Kaczmarczyk, 2008; Moos and Thanoukos, 2009), which features side burners and a removable top. Removing the entire top (see Figure 9.2) allows the furnace to be charged with fewer bucket loads of scrap. This can reduce charging times by up to 80%, which keeps the furnace refractories from

FIGURE 9.2 Round melting furnace. (From Boeckenhauer, K. and Kaczmarczyk, T., Modern furnaces for aluminum scrap recycling, http://www.secowarwick.com/assets/Documents/Articles/Aluminium/MODERN-FURNACES-FOR-ALUMINIUM-SCRAP-RECYCLING-AP.pdf, 2008. With permission; photo courtesy of T. Kaczmarczyk.)

getting cold. The reduced number of charging steps also improves safety by reducing worker exposure to the furnace environment. Round furnaces are easier to seal than other furnace types and typically have higher melting rates per unit area.

Dry-hearth furnaces solve many of the problems associated with wet-hearth melting (Newman, 2010). A particular advantage is the reduction in melt loss caused by the use of different types of flame over the solid scrap and the metal bath. Convective flames are used in the dry-hearth area to maximize heat transfer to the solid scrap; flat luminous flames work better to heat the molten metal. As a result, dry-hearth furnaces are the most popular approach for melting large or bulky scrap.

However, dry-hearth furnaces also have limitations. Refractories in the elevated portion of the hearth are more prone to breakage by the solid scrap being dumped on them without a cushion. The lack of intimate contact between combustion gases and the charge means that thermal efficiencies are still low (30%–35%). Scrap size is still a limitation as well, since smaller scrap piled on the hearth is less likely to stay there. Melt loss can be higher than 10%, particularly for thin scrap (Grayson, 2012).

The next step in the progression is the *stack* melter shown in Figure 9.3. Here scrap is input directly into the exhaust stack, forcing the exhaust gas to go through the scrap as it leaves the furnace. As the heated scrap descends to the sloping hearth, additional burners melt it, causing it to flow into the molten bath. This in turn allows more scrap to descend to the hearth, creating a semicontinuous melting operation.

Stack melting has several advantages over traditional single-chamber melting (Kendzora, 2000; Kennedy, 2001). The most significant is improved efficiency. With most of the heat removed from the exhaust gas (exit temperatures are reduced to 500°C or less), energy savings of 30%–50% or more are common (White, 2011). The preheating

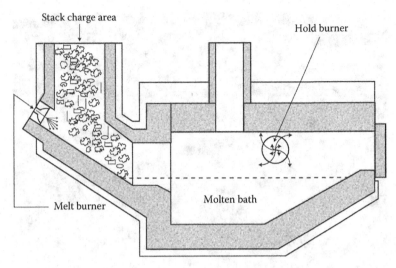

FIGURE 9.3 Stack melter. (From McKenna, J.P. and Wisdom, A., *Die Cast. Eng.*, 41(6), 64, June 1997. With permission.)

of the scrap by the exhaust gas also reduces melting time on the hearth, improving furnace productivity. The elimination of the open charging well reduces melt loss by 80% or more and minimizes emissions. Stack melting is not suitable for very large scrap, but a stack can be combined with a dry hearth to accommodate a range of charge materials. These advantages often justify the additional investment required for a stack melter.

Although a considerable improvement over simpler single-chamber furnaces, stack melting has a few disadvantages of its own. The foremost is controllability; scrap descends as quickly as it melts, which makes slowing down the melting rate difficult. The pressure of the scrap on the stack limits the depth to which it can be stacked, and this in turn limits the length of time that it can be preheated. Uneven heating is also a concern if the charged scrap has a range of sizes. The solution introduced in the early 1970s was the *tower melter*. The tower through which exhaust gas flows consists of several chambers separated by cast iron bars. Scrap is introduced into the top chamber while melting is completed in the furnace hearth below. When metal is tapped from the furnace, the bars on each chamber are sequentially released, and the scrap falls into the next chamber. Very large scrap can be charged directly onto the dry hearth in the main furnace chamber as needed. The design allows a greater amount of scrap to be held in the column, improving heat recovery and ultimately furnace efficiency (values of 60%–77% are claimed). The lack of direct contact between the flame and the scrap also helps reduce melt loss. Figure 9.4 shows the latest version of the Ecomelt furnace (Niedermair, 2009, 2011), in which furnace off-gas fed through the tower decoats high-organic scrap. The gas coming out of the tower is then burned to provide supplemental heat to the system. The Ecomelt includes a separate melting chamber similar to the sidewell furnace, along with electromagnetic pumps to circulate metal (see previous chapter). This allows processing of high-organic scrap and other materials that would normally be shunned by remelters.

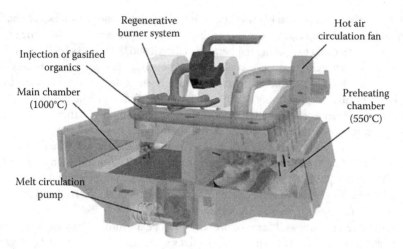

FIGURE 9.4 Ecomelt multichamber melting furnace. (From Hertwich Engineering Multi Chamber Furnace (type Ecomelt), 14 August 2013, http://www.hertwich.com/index.php?id = casting_furnaces. With permission.)

However, tower melters have higher capital costs than other furnaces, and the number of moving parts is a potential maintenance headache. The *footprint* of stack and tower melters is also substantial, and this makes it difficult to fit them into existing shop floors. As a result, the use of tower melting furnaces is uncommon and largely limited to smaller melting operations.

MULTIPLE-CHAMBER FURNACES

Figure 9.5 illustrates the *sidewell* melting furnace, which has been used for several decades for the melting of light scrap (Feese and Lesin, 2009). The sidewell furnace was designed to eliminate direct interaction between combustion gases and solid scrap, which increased melt loss and dross generation. Instead, only the molten metal in the *clean chamber* on the left is heated, using high-luminosity flames (see previous chapter).

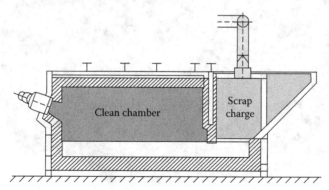

FIGURE 9.5 Cross section of a simple sidewell furnace. (From de Groot, J. and Migchielsen, J., in *Aluminium Cast House Technology—Eighth Australasian Conference*, Whiteley, P.R., Ed., TMS-AIME, Warrendale, PA, p. 57, 2003. With permission.)

The superheated molten metal flows underneath a baffle and contacts solid scrap charged to the sidewell, heating and ultimately melting it. When the charge is melted, the molten metal is then reheated to pouring temperature, tapped, and the process is repeated.

Sidewell furnaces are an effective way to melt light scrap such as UBCs with minimal melt loss. However, improvements have been desired ever since their introduction. Because no preheating of scrap is provided, fuel efficiencies are poor. The use of tactics such as regeneration or air enrichment to improve efficiency has been a common response (Stewart, 2002). A bigger problem has been the slow kinetics of melting (Migchielsen and Gräb, 2007), limited by the rate at which metal can flow under the baffle to contact the scrap in the well. The metal pumping devices described in the previous chapter were designed largely to improve metal flow rates around the baffle and increase melting rates (Henderson et al., 2001). Both mechanical and electromagnetic pumps are widely used.

Even though the sidewell furnace prevents direct exposure of the scrap to combustion flames, the finely divided material oxidizes sufficiently during heating to generate a significant amount of dross. Until the 1990s, the only response to this was to use salt as a flux (Foseco, 2002), which minimized metal loss to the dross but generated an environmentally undesirable waste product. As a result, a means was sought to submerge the pieces of scrap in the molten metal in the sidewell while they melted, which would minimize oxidization of the scrap and eliminate the need for fluxing. Figure 9.6 illustrates a vortex system (Iribarren, 2008; Pyrotek, 2005; Starczewski, 2012). Furnaces using this approach have an additional chamber into which molten metal is pumped at high velocity. The design of the chamber causes the molten metal to form a vortex, in which small pieces of scrap (chips, UBC shreds, wire chops) are

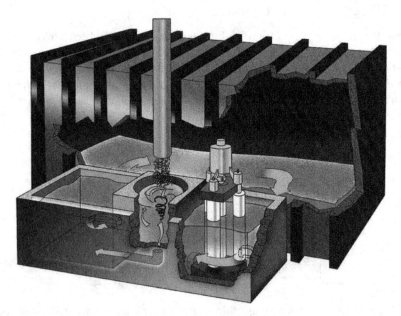

FIGURE 9.6 LOTUSS (Low Turbulence Scrap Submergence) system. (From Metaullics Systems, http://www.metaullics.com/lotuss.html. With permission.)

fed. Output from the vortex goes to the sidewell, where it finishes melting before moving back to the main chamber. Devices of this type have been available for over 30 years and have been adopted worldwide. Recent improvements have allowed this approach to be used in larger furnaces and have increased melt capacity.

The problem of melting small scrap becomes greater if coating material and other organics are still present when the material is charged. For scrap of a consistent type (such as UBCs), a preliminary decoating step (see Chapter 5) is justified, but mixed scrap loads are often charged directly to the melting furnace without prior decoating. This leads to higher melt losses, as carbon and oxygen in the coatings react with the aluminum to generate alumina and aluminum carbide. The gases given off by decomposing coatings are also an environmental concern. Figure 9.7 presents a solution to this: the twin-chamber furnace developed in Germany (Migchielsen and

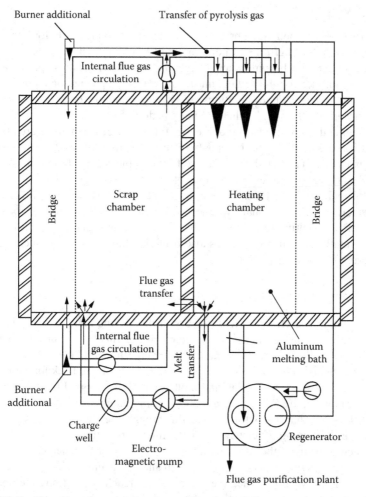

FIGURE 9.7 Top view of a twin-chamber melting furnace for contaminated scrap. (From Meyer, H.J., *Ind. Heat.*, 70(2), 30, February 2003. With permission.)

Gräb, 2007; Valder, 2011). The furnace features a scrap chamber, which is effectively an enlarged sidewell, and a heating chamber where molten metal is superheated. An electromagnetic pump circulates metal between the two chambers; regenerating burners improve thermal efficiency. The most significant advance in the furnace is a recirculating flue gas circuit, in which gases given off by decomposing coatings are burned to recover additional energy and minimize emissions. Temperatures at the top of the scrap chamber are controlled at 500°C to maximize pyrolysis rates. The nonoxidizing atmosphere in the furnace helps minimize melt loss, and this in turn allows operation without salt. If the quality of the scrap supply is inconsistent enough, these furnaces are a good choice for environmentally viable melting. Scrap with up to 6% contamination can be successfully processed. For a steady supply of coated scrap, separate decoaters are a more competitive choice.

SMALL-VOLUME MELTERS

The furnaces described in the previous sections are used primarily for large volume applications (>1000 kg/h melt rate). For smaller-volume melting operations, minimizing capital outlay becomes increasingly important, even at the expense of energy efficiency. Smaller-volume melters must also be easily operated on an on–off basis, rather than the steady-state operations used by larger melting operations. This means that a furnace with a smaller thermal mass is desirable, as is one that can be completely emptied and started up cold. Electric furnaces (see Chapter 10) are often used for applications of this type, but gas-fired furnaces are also used where electricity is too expensive. The small size of these furnaces means that liquid fuels (fuel oil, diesel, LPG) are sometimes used instead of natural gas.

Fossil-fuel crucible furnaces are available in stationary or tilting models (Nabertherm, 2012); stationary furnaces can have either removable or fixed (bale-out) crucibles. Melting rates are usually <0.5 ton/h, and fuel efficiencies less than 20%. A small dry-hearth melter has also been developed, useful for melting wet or oily scrap (Thermtronix, 2009).

ROTARY FURNACES

For highly oxidized scrap or dross, the use of flux during melting is a requirement. Separating the resulting salt slag from the metal is difficult without sufficient agitation. To deal with this problem, the tilting rotary furnace (TRF) shown in Figure 9.8 was developed. The rotary furnace is a sloping tube, tilted back for charging and firing and tilted forward for slag and molten metal discharge. Firing occurs at the bottom (Maiwald, 2009; Zhou et al., 2006), using natural gas or occasionally fuel oil. As the hot exhaust gas travels upward through the furnace, it passes through the charge, improving efficiency and reducing melting times. The gas also heats the refractory walls of the furnace; this allows heating of the charge by conduction from the walls as the furnace rotates (Jepson and van Kampen, 2005), as well as by radiation and convection. When the contents have reached the desired temperature, the furnace is tilted and the metal and salt slag are separately poured off. Rotation of the furnace is slow during the early stages of the melting process to prevent damage

FIGURE 9.8   View of a rotary melting furnace. (From Boeckenhauer, K. and Kaczmarczyk, T., Modern furnaces for aluminum scrap recycling, http://www.secowarwick.com/assets/ Documents/Articles/Aluminium/MODERN-FURNACES-FOR-ALUMINIUM-SCRAP-RECYCLING-AP.pdf, 2008. With permission; photo courtesy of T. Kaczmarczyk.)

to the refractories from large pieces of scrap or dross; as a molten pool begins to form, the speed is increased to roughly 7–8 m/s at the inner surface. Melting times vary with furnace size and loading but are usually shorter than in a comparable wet-hearth furnace.

Rotary furnaces are faster and more efficient than ordinary reverberatory furnaces (Paitoni and Benedini, 2004). Because they are emptied completely after each tap, they are more suitable than stationary furnaces for aluminum producers who change alloys frequently (Boeckenhauer and Kaczmarczyk, 2008). However, they are also more expensive to install per unit capacity and more difficult to maintain; labor costs are also higher. The need for salt also presents challenges, because salt mixtures with a lower melting point than that of the metal feature high levels of expensive KCl. As a result, TRFs are generally best suited for melting dross and other oxidized scrap (see Chapter 13) and for smaller-size scrap like UBCs (Hall, 2008). However, the smallest material, such as clean punchings and turnings, are now better suited to a vortexing furnace, where they can be melted in a salt-free environment.

HOLDING AND DOSING FURNACES

In smaller melting facilities, molten metal tapped from a melting furnace is usually transferred to a ladle prior to casting. Larger facilities often employ a holding furnace (Whipple, 2005), where the temperature and composition of the metal can be adjusted. This allows the melting furnace to be used mostly for its intended purpose, increasing productivity. The use of a holding furnace also makes it possible to operate

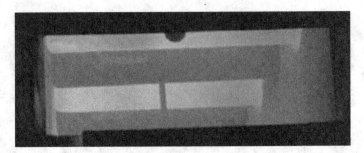

FIGURE 9.9   Inside of a radiant-tube furnace. (From Sandvik Materials Technology, on-line brochure, "Gas Heated Furnaces". With permission.)

the melting furnace under conditions that most favor heating of solid scrap (reducing conditions, convective heat transfer), improving the efficiency and melting rate.

Many different holding furnaces are available, and both gas-fired and electric furnaces are widely used (Butler, 2006; Nabertherm, 2012; Schaefer, 2008). Figure 9.9 illustrates a well-known concept in gas-fired holding furnaces: a *radiant-tube* furnace (Flamme et al., 2011; Wunning, 2002). In this furnace, products of combustion are passed through ceramic tubes mounted in the furnace; as the tubes heat up, they radiate energy to the molten metal below. The use of a tube separates the combustion gases from the metal, limiting oxidation and disturbance and reducing melt loss. Electrical resistance elements mounted in the roof have the same effect. Reverberatory furnaces operating under luminous flame conditions are less expensive and are also common.

RECOMMENDED READING

Boeckenhauer, K. and Kaczmarczyk, T., Modern furnaces for aluminum scrap recycling, November 18, 2008. http://www.secowarwick.com/assets/Documents/Articles/Aluminium/ MODERN-FURNACES-FOR-ALUMINIUM-SCRAP-RECYCLING-AP.pdf.
Butler, W.A., Melting and holding furnaces for die casting, *Die Casting Engineer*, (3), 28, March 2006.
Pyrotek, *Improving Performance in Casthouse Furnaces*, August 29, 2005. http:// www.pyrotek.info/furnace_operations.
Schaefer Furnaces, *Furnaces for Melting and Holding Aluminum*, February 25, 2008. http:// www.fwschaefert.com/document/CompleteBrochure2008.pdf.
Schmitz, C., *Handbook of Aluminium Recycling*, Vulkan-Verlag, Essen, Germany, 2006, p. 106.

REFERENCES

Barry, S., Rodriguez, F., and Gil, O., CVG venalum—Design of a 55 t tilting melting furnace, in *Light Metals 2009*, Bearne, G., Ed., TMS-AIME, Warrendale, PA, 2009, p. 609.
Boeckenhauer, K. and Kaczmarczyk, T., Modern furnaces for aluminum scrap recycling, November 18, 2008. http://www.secowarwick.com/assets/Documents/Articles/Aluminium/ MODERN-FURNACES-FOR-ALUMINIUM-SCRAP-RECYCLING-AP.pdf.
Brooks, C.L., *Basic Principles of Aluminum Melting, Metal Preparation, and Molten Metal Handling*, Reynolds Metals Co., Richmond, VA, 1970, p. 55.

Butler, W.A., Melting and holding furnaces for die casting, *Die Casting Engineer*, 50(3), 28, March 2006.

Emes, C.B., Improvements in metal quality and operating efficiency through furnace design, presented at *International Melt Quality Workshop*, Madrid, Spain, October 25, 2001. http://www.pyrotek.info/documents/techpapers/Metal-Quality-thru-Furnace-Design-CBEmes.pdf.

EMP Technologies, Raising the level of electromagnetic pumping, *Alum. Int. Today*, 19(3), 24, May/June 2007.

Feese, J. and Lesin, F., Reducing metal loss in side well charged melters with Invisiflame® burner technology, in *Light Metals 2009*, Bearne, G., Ed., TMS-AIME, Warrendale, PA, 2009, p. 721.

Flamme, M., Milani, A., Wünning, J.G., Blasiak, W., Yang, W., Szewczyk, D., Sudo, J., and Mochida, S., Radiant tube burners, Chapter 24 in *Industrial Combustion Testing*, Baukal, C.E., Ed., CRC Press, Boca Raton, FL, 2011, p. 487.

Foseco, *Development, Evaluation, and Application of Granular and Powder Fluxes in Transfer Ladles, Crucible and Reverberatory Furnaces*, April 30, 2002. http://www.foseco.com. tr/tr/downloads/FoundryPractice/237-02_Development_evaluation_appli_of_granular_flu-1.pdf.

Grayson, J.T., Cost savings in the cast house: Optimizing furnace operation, staff training and associated variables, Pyrotek MCR Group, March 21, 2012. http://www. improvingperformance.com/papers_post_event/Case_Study_5_Cost_savings_in_the_casthouse_furnace_practices_Jim_Grayson_Pyrotek_MCR_Group.pdf.

de Groot, J. and Migchielsen, J. Multi chamber melting furnaces for recycling of aluminum scrap, in *Aluminium Cast House Technology—Eighth Australasian Conference*, Whiteley, P.R., Ed., TMS-AIME, Warrendale, PA, 2003, p. 57.

Hall, C., Tilt rotary replacing traditional reverberatory melting, *Aluminium*, 84, 38, 2008.

Henderson, R.S., Neff, D.V., and Vild, C.T., Recent advancements in gas injection technology using molten metal pumps, in *Light Metals 2001*, Anjier, J., Ed., TMS-AIME, Warrendale, PA, 2001, p. 1033.

Hertwich Engineering, Multi Chamber Furnace (type Ecomelt), 14 August 2013, http:// www.hertwich.com/index.php?id=casting_furnaces.

Iribarren, O., *Aluminium Chips Recycling: High Efficiency and Recovery on Aluminium Foundries*, March 11, 2008. http://www.aluplanet.com/documenti/InfoAlluminio/InfoInsertecENG.pdf.

Jepson, S. and van Kampen, P., Oxygen–enhanced combustion provides advantages in Al-melting furnaces, *Ind. Heat.*, 72(6), 29, June 2005.

Kearney, A., Reverberatory furnaces and crucible furnaces, in *Metals Handbook, Vol. 15: Casting*, 9th edn., ASM International, Materials Park, OH, 1988, p. 374.

Kendzora, S., Melting furnaces in die casting, *Die Casting Engineer*, 44(5), 28, September/October 2000.

Kennedy, S., Aluminum melting and metal quality processing technology for continuous high quality castings, in *Sixth International AFS Conference: Molten Aluminum Processing*, AFS, Des Plaines, IL, 2001.

Maiwald, D., Advanced control of a rotary drum furnace in a secondary smelter, in *Light Metals 2009*, Bearne, G., Ed., TMS-AIME, Warrendale, PA, 2009, p. 615.

McKenna, J.P. and Wisdom, A. Energy savings in gas-fired aluminum reverberatory furnaces, *Die Cast. Eng.*, 41(6), 64, June 1997.

Metaullics Systems, http://www.metaullics.com/lotuss.html.

Meyer, H.J. Salt-free aluminum scrap melting in twin-chamber furnace, *Ind. Heat.*, 70(2), 30, February 2003.

Migchielsen, J. and Gräb, H.-W., Newest developments of multi chamber melting furnaces, in *Light Metals 2007*, Sørlie, M., Ed., TMS-AIME, Warrendale, PA, 2007, p. 609.

Moos, O. and Thanoukos, E., Installation of a new 135 tonne RTC melting furnace, *Aluminium*, 85, 69, 2009.

Nabertherm, *Foundry*, June 13, 2012. http://www.nabertherm.com/produkte/giesserei/giesserei_english.pdf.

Newman, P., Dry hearth melting furnaces, *Mater. Sci. Forum*, 630, 103, 2010.

Niedermair, F., Latest trends in scrap remelting, presented at *10th OEA International Aluminium Recycling Congress*, Berlin, Germany, March 02, 2009.

Niedermair, F., Benchmark remelt plant for Hai, Romania, automated billet production dross—Quo vadis, presented at *12th OEA International Aluminium Recycling Congress*, Vienna, Austria, February 21, 2011.

Paitoni, C. and Benedini, L., Rotary smelting furnace, *Diecast. Technol.*, (31), 52, September 2004.

Pyrotek, *Improving Performance in Casthouse Furnaces*, August 29, 2005. www.pyrotek.info/furnace_operations.

Sandvik Materials Technology, on-line brochure, "Gas Heated Furnaces".

Schaefer Furnaces, *Furnaces for Melting and Holding Aluminum*, February 25, 2008. www.fwschaefert.com/document/CompleteBrochure2008.pdf.

Schmitz, C., *Handbook of Aluminium Recycling*, Vulkan-Verlag, Essen, Germany, 2006, p. 106.

Starczewski, R., Developments in scrap submergence technology for light gauge scrap and alloy charging (LOTUSS® technology), presented at *20th Metal Bulletin Recycled Aluminium Conference*, Salzburg, Austria, November 20, 2012.

Stewart, D.L., Jr., Aluminum melting technology—Current trends and future opportunities, in *Light Metals 2002*, Schneider, W., Ed., TMS-AIME, Warrendale, PA, 2002, p. 719.

Thermtronix, Non-crucible aluminum melting furnace, June 30, 2009. http://www.thermtronix.com/PAGE9.HTM.

Valder, G., Technology of fuel-fired melting furnaces at Otto Junker, *Aluminium*, 87(3), 42, 2011.

Whipple, D.F., Improved control for aluminum holding furnaces, *Bloom Engineering*, May 20, 2005. https://www.bloomeng.com/uploads/IMPROVEDCONTROLALUMHOLDING.pdf.

White, D.W., Furnaces designed for fuel efficiency, in *Light Metals 2011*, Lindsay, S.J., Ed., TMS-AIME, Warrendale, PA, 2011, p. 1169.

Wunning, J.G., Ceramic radiant tubes extend performance limits, *Ind. Heat.*, 69(3), 73, March 2002.

Zhou, B., Yang, Y., Reuter, M.A., and Boin, U.M.J., Modelling of aluminium scrap melting in a rotary furnace, *Miner. Eng.*, 19, 299, 2006.

10 Electric Furnace Melting

As illustrated in previous chapters, most aluminum scrap is melted in furnaces heated by the combustion of a fossil fuel—natural gas, oil, or perhaps coal. However, the use of fossil fuels as an energy source has disadvantages (Butler, 2006):

- Poor fuel efficiency
- Required ventilation to handle the combustion products
- Dross generation from the interaction of combustion products with the molten metal
- Melt contamination from the interaction of combustion products with the molten metal
- High surface temperature in the metal, which encourages hydrogen absorption
- Refractory degradation in the roof and sidewalls
- Metal contamination from entrained dross circulated by metal pumps and electromagnetic stirring

Electricity has been used as an alternate energy source for melting aluminum scrap for some time. The first electric furnace for melting scrap was installed in the United States in 1918 (Anderson, 1987) and appeared in Europe soon afterward. These furnaces have been in use ever since and have evolved over time. The most important innovation was the introduction of induction furnaces for melting in the 1930s. These have largely displaced the resistance and arc furnaces originally used.

Electric furnaces have important advantages over fossil-fuel furnaces for melting aluminum scrap. The most important of these is cleaner metal. Because there are no combustion products in an electric furnace environment, dross generation is much less in an electric furnace (Lessiter, 1997), as is gas pickup. As a result, melt losses are lower (Groteke, 1997; Hentschel and Feldmann, 1982; Kreysa, 1991; MacIntosh, 1983), and metal purity is improved. The stirring motion of an induction furnace minimizes temperature gradients within the melt (Heine and Gorss, 1991), improving consistency. Electric furnaces are generally more efficient than gas- or oil-fired furnaces (Fishman, 2002; Schifo and Radia, 2004), especially in smaller sizes. The lack of fuel combustion means that there is little or no off-gas, which greatly reduces environmental concerns. In addition, electric furnaces are less noisy.

However, there are also disadvantages. The stirring motion that improves homogeneity can also prevent dross from separating, increasing the risk of inclusions (Whiteley, 2011). While electric furnaces are more efficient, electricity is often a more expensive form of energy than fossil fuels, erasing the cost advantage (Fishman, 2002; Nealon, 2011); recent declines in the price of natural gas have made electricity an even less desirable option. Electric furnaces tend to have higher capital costs than fossil-fuel furnaces of equivalent capacity (Hellsing and Tallbäck, 1982; MacIntosh, 1983). Most importantly, electric furnaces have difficulty matching the melting capacity of large-scale reverberatory furnaces (Butler, 2006; Groteke, 1997).

As a result, electric furnaces are most often found in small-volume operations such as foundries. Less than 5% of the total aluminum melted in the United States is melted in electric furnaces (Schifo and Radia, 2004). However, changing environmental restrictions and the need to produce cleaner metal in a melting furnace may encourage greater use of electric furnaces in the future.

The electric furnaces used in aluminum melting can be classified in several ways. The most important is the type of electrical heat generation, which is by *induction* or *resistance*. Induction furnaces can be further divided into *channel* and *coreless* units. A second means of classification is the function of the furnace. Some electric furnaces operate as melting units, others as holding furnaces, and some as both. The choice of frequency is another means of classifying electric furnaces, as is the means by which molten metal is removed (in a separate crucible, by hand-dipping, or by being tapped to a ladle). In this chapter, electric furnaces will be sorted first into induction and resistance furnaces and then by the other classification methods.

INDUCTION FURNACES

In Chapter 5, the concept of an eddy current was introduced as a technology for separating aluminum from other nonmagnetic metals in a scrap stream. An eddy current is the electric current induced in a conductor by the magnetic field associated with an adjacent magnet or electric current. In an eddy current separator, powerful permanent magnets are typically used to generate this field. Passing a piece of scrap over a series of magnets with alternating polarities generates Lorentz forces, which repeatedly flip the piece of scrap and cause it to veer from its original path.

If the scrap piece (or any other conductor) remains in the magnetic field generated by alternating current for a period of time, its resistance to the reversing eddy current generated by the field will cause it to heat up. This is called induction heating and is used for several metallurgical applications. The industrial use of induction heating for melting purposes dates from the 1930s (Robiette, 1972), and today is used for most nonferrous metals, as well as irons and steels. Two general approaches have been taken in the design of induction-based melting furnaces. In the first, solid metal is directly heated and melted by eddy currents in a coreless furnace. In the second, molten metal is heated to a high temperature in the bottom of a channel induction furnace; the superheated molten metal then melts added scrap or ingot charged to the top of the furnace.

CORELESS FURNACES

Figure 10.1 shows the construction of a typical coreless induction furnace (Bala, 2005; Perkul, 2008). The induction coil surrounding the outside of the lining containing the metal charge consists of water-cooled copper tubing, embedded in

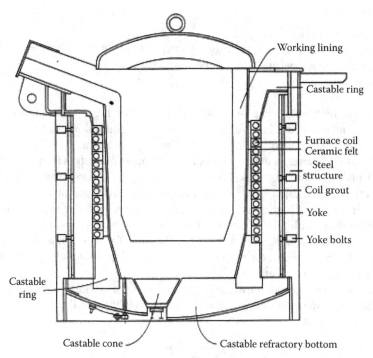

FIGURE 10.1 Construction of a coreless induction furnace. (From Perkul, R.Y., in *Metals Handbook, Vol. 15: Casting*, 10th edn., ASM International, Materials Park, OH, 2008. With permission.)

a ceramic grout to reduce corrosion and provide additional support. In furnaces with removable crucibles, the coil will reach nearly to the top of the furnace. In tiltable furnaces like that shown here, room is left for a pouring spout. The magnetic field generated by the coil is equally strong on both sides; to prevent the stray flux from heating the outer furnace shell, a series of vertical laminations of transformer iron known collectively as the yoke is located outside the coil. In addition to tiltable furnaces like this, coreless induction furnaces are available that can be tapped from the bottom, along with removable crucible furnaces (Lessiter, 1997).

The biggest decision in specifying a coreless induction furnace is the choice of operating frequency. Until the 1970s, there was no such choice; virtually all coreless units operated at *mains frequency,* 50–60 Hz. This had several consequences. The first stems from the impact of furnace frequency on the penetration depth of the magnetic field. Robiette (1972) describes penetration depth using the expression

$$p = \frac{1}{2\pi}\sqrt{\frac{\rho}{\mu f}} \tag{10.1}$$

where
  p is the penetration depth in cm
  ρ is the resistance of the conductor affected by the field (i.e., the furnace charge)
  μ is the magnetic permeability of the conductor
  f is the frequency

For a given material, p is proportional to $1/f^{0.5}$. The high penetration of mains frequency means that a large-diameter furnace is required to eliminate bucking or canceling of the field (Hipple, 1982). However, practical considerations limited the coil diameter to a maximum of 3.65 m outer diameter (Heine and Gorss, 1991; Knödler, 1986). As a result, mains frequency induction furnaces often have lower efficiencies, especially during melting when the resistance of the charge is greater (and p is higher as a result). Furthermore, Knödler suggests that the minimum size of aluminum scrap (or ingot) that can be efficiently melted in a coreless furnace is $3.5p$. This means that the use of mains-frequency induction furnaces for melting is best reserved to ingot and large-diameter scrap such as casting runners and gating.

A second consequence of mains-frequency furnace operation is the impact of frequency on stirring action in the melt. Shuichi (1988) calculates *stirring action* with the expression

$$F = \alpha \frac{P}{\sqrt{fDH}} \tag{10.2}$$

where
  α is a constant determined by the metal being melted
  P is the power input to the furnace
  D is the inside crucible diameter
  H is the coil height

Lower frequencies mean increased stirring. This stirring theoretically makes it easier to melt thin-gauge scrap (Perkul, 2008), by rapidly drawing the scrap under the melt surface. However, stirring is also more likely to hurt metal quality, by increasing melt contact with the atmosphere and increasing hydrogen absorption. The increased stirring also raises the height of the *meniscus* or the dome of molten metal that forms above the melt. A meniscus that gets too high can result in metal spillage. Preventing this requires (a) turning down the power (and reducing furnace melt capacity) or (b) increasing the distance by which the crucible wall extends above the coil, which creates thermal gradients within the melt. As a result, the use of mains-frequency coreless induction furnaces for melting scrap is not widespread. When used for this purpose, they are generally operated with a hot heel of molten metal (Smith and Hayes, 1992), to which scrap is added. This reduces the resistance of the charge and allows the stirring action to transfer heat to the solid scrap more effectively.

In the 1970s, the development of solid-state frequency converters (thyristorized or oscillating circuit) made possible *medium-frequency* (150–600 Hz)

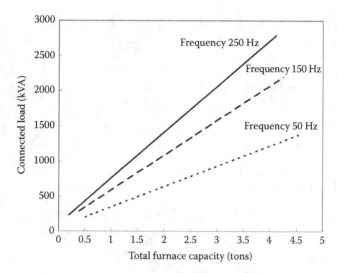

FIGURE 10.2 Coreless induction furnace load vs. capacity for different frequencies. (From Knödler, G., *Cast. Plant + Technol.*, (1), 22, January 1986. With permission.)

furnace operation (Knödler, 1986). Medium-frequency operation reduces the penetration depth generated by the coil, compared with mains-frequency operation. Higher frequency operation also decreases the stirring and thus the meniscus height. This in turn increases the power density that can be used for a furnace of a given size (Goyal, 1985), as Figure 10.2 illustrates. This power increase means reduced melting times for a given charge, and the reduction in furnace size made possible by the medium-frequency operation reduces capital costs and improves efficiency (Hipple, 1982). The reduced stirring and shorter melting times translate into lower melt losses than mains-frequency melting, and the reduced penetration depth means that smaller pieces of scrap can be melted (Kreysa, 1991). Another advantage of medium-frequency melting is that a cold charge can be melted more easily, while mains-frequency furnaces generally use a molten heel (Heine and Gorss, 1991).

The optimal frequency of coreless induction furnace to be used for melting scrap depends on the capacity, the size of material to be melted, and equipment and electricity costs. Figure 10.3 shows an approach suggested by Hipple (1982); the greater the capacity (and the larger the furnace), the lower the frequency. A second approach can be obtained from the work of Rowan (1987), who published a frequency selection chart, based on experience with a variety of foundries. The published chart recommends an optimum frequency for a given furnace capacity, with a margin of uncertainty called the *green zone*. The published chart is primarily for ferrous melting furnaces but can be adapted for frequency recommendation for aluminum melting as well. From Rowan's results, the following expression can be obtained:

$$\log y = (-0.6056 \log x + 4.409) \pm 0.3$$

(10.3)

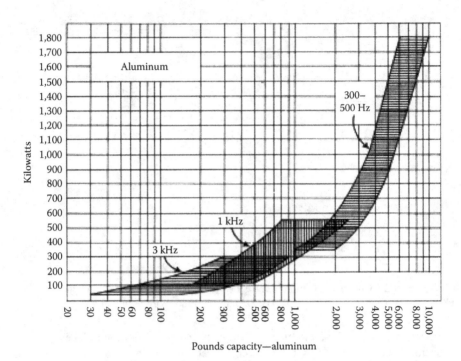

FIGURE 10.3 Coreless induction furnace load vs. capacity, showing optimal frequency ranges. (From Hipple, W.F., *Modern Cast.*, 72(8), 39, August 1982. With permission.)

where
  y is the optimal frequency
  x is the capacity of the furnace in kg

For a furnace with a capacity of 3000 lb (1361 kg), the recommended frequency from the expression would be 324 Hz, at the low end of what Figure 10.3 suggests. A furnace with a capacity of 8000 lb (3629 kg) would have a recommended frequency of 111 Hz, well below that suggested by Hipple (1982).

Figure 10.4 shows a common power supply schematic for medium-frequency furnaces (Perkul, 2008). The equipment shown here is a parallel inverter, in which the capacitor bank is connected in parallel with the induction coil; series inverters are also used. The lack of tap changes and capacitor contactors means that supply line surges do not occur, which pleases power companies; it also means reduced maintenance costs. For furnaces with power ratings above 1250 kW, a step-down transformer is normally used; smaller furnaces are more likely to use a 480 V distribution line backed by a disconnecting device. Power supply functions are increasingly automated, using programmable controllers and software packages that can specify appropriate sequences for different types of furnace operation (cold start vs. hot start, different types of metal input, batch vs. tap-to-tap, etc.).

Coreless induction furnaces are operated in one of two basic methods (Perkul, 2008). The first is a batch operation, in which the furnace is completely emptied

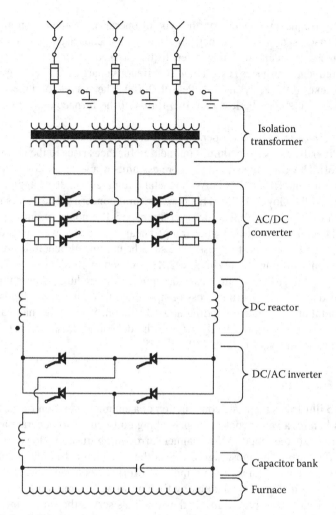

FIGURE 10.4 Solid-state frequency conversion schematic for medium-frequency induction furnaces. (From Hipple, W.F., *Modern Cast.*, 72(8), 39, August 1982. With permission.)

after every heat. The second is a tap-and-charge operation, in which the furnace is operated semicontinuously, with about 50% of its contents removed after each heat. The more consistent temperature of a tap-and-charge operation ensures longer refractory life; however, batch operation has a higher electrical efficiency (Kreysa, 1991). While tap-to-tap times are longer in batch melting, the frequency of tapping is reduced, ultimately maximizing furnace utilization. In addition, the ability to operate medium-frequency coreless furnaces with a completely solid charge means that some oily or wet scrap can be used in batch operation, as long as no liquid heel is left after tapping. (The oil and water are vaporized during heatup.) This practice remains undesirable, however, and it is recommended that the fraction of wet scrap in a charge be limited to one-fifth or less of the total (Groteke, 1997).

Charging frequency depends on the type of operation (batch or tap-and-charge) and the furnace size and frequency. The lower melt capacity of mains-frequency coreless furnaces means that they need to be charged less often; the appetite of medium-frequency furnaces is such that a continuous charging device may be appropriate (Groteke, 1997; Kreysa, 1991; Perkul, 2008). The size of solid pieces charged to a coreless unit should be less than two-thirds of the furnace diameter to prevent bridging.

A peculiar concern in the operation of coreless furnaces is the disposition of dross. Ordinarily, dross from aluminum melting furnaces rises to the top, where it is skimmed off. However, the stirring in coreless units draws dross down instead into the melt (Smith and Hayes, 1992). As the metal circulates, the dross agglomerates on the furnace walls (Goyal, 1985). This poses several concerns. The first is refractory penetration (Shuichi, 1988). The dross contains small amount of entrapped metal, which will become superheated when trapped near the crucible wall (where the magnetic field from the coil is strongest). The superheated metal penetrates refractories easily, shortening lining life (Whiteley, 2011). A second problem is the increase of lining thickness resulting from dross adsorption. This reduces power input to the melt. A third possible concern is dross being eroded from these side deposits by circulating metal and ultimately winding up in the metal. Special mechanical scrapers have been designed to periodically remove the deposited dross without disturbing the molten metal surface (Perkul, 2008).

CHANNEL FURNACES

Figure 10.5 illustrates the basic construction of a channel induction furnace (Perkul, 2008). As before, a water-cooled copper-tubing coil is used to generate a magnetic field, which heats the metal in the channel surrounding the coil. However, there are several differences. The most significant is that the channel is placed outside the coil, whereas the metal in a coreless induction furnace is placed inside the coil. To redirect the magnetic field generated by the coil to act entirely on the metal in the channel, an iron core is placed inside the coil. This serves the same function as the iron yoke placed outside the coil in a coreless unit.

A bigger difference between the channel and coreless induction furnaces is that the amount of metal in the channel surrounding the coil is much smaller than the amount of metal being acted on by the field in a coreless unit. As a result, channel furnaces heat metal by an indirect method. Metal flowing through the channel is superheated by the field and moved along by the same forces that cause bath circulation in coreless units. As the superheated metal exits the channel into the main body of the furnace, it transfers heat to the rest of the melt. (This can be seen as an electric-furnace analog to sidewell melting furnaces heated by fossil-fuel combustion.) Scrap or ingot fed to the furnace is melted by this indirect approach, rather than going through the channel.

Nearly all channel furnaces operate on mains frequency. Because of this, the power density is limited, making the direct melting of scrap impossible. This in turn means that a heel of molten metal must be maintained in the furnace (Perkul, 2008). This reduces furnace flexibility and makes alloy changes

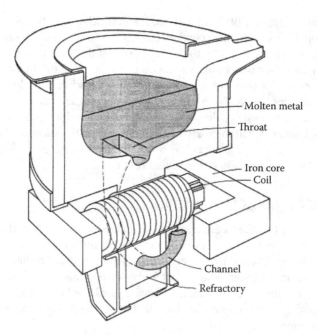

Molten metal

Throat

Iron core
Coil

Channel

Refractory

FIGURE 10.5 Construction of a channel induction furnace. (From Perkul, R.Y., Induction furnaces, in *Metals Handbook, Vol. 15: Casting,* 10th edn., ASM International, Materials Park, OH, p. 108, 2008. With permission.)

difficult. As a result, channel furnaces are used more often as holding furnaces rather than as melting units and typically for larger operations. However, for these applications, channel furnaces have advantages. Their efficiency is typically 70%–75%, compared with the 50%–55% efficiencies of coreless furnaces (Knödler, 1986). A second advantage is the reduced agitation in channel furnaces. This minimizes dross formation compared with coreless units and reduces gas pickup. However, the reduced turbulence also makes it more difficult to charge low-bulk scrap, such as cans, shredded material, and trimmings (Kreysa, 1991). These materials must be baled or otherwise densified before charging to reduce melt loss.

Until the early 1980s, a second limitation on channel furnace capacity was the induction coil itself, which was air cooled (Hellsing and Tällback, 1982; Knödler, 1986). This limited inductors to power ratings of 250 kW and less. Even so, inductor failure is a frequent problem with these furnaces (Whiteley, 2011). The development of water-cooled inductors allowed power ratings to increase to 1000 kW or higher. Using multiple induction loops on a given furnace also increases furnace capability. However, this comes at a price; channel furnaces are more expensive per unit melting capacity than coreless units (and much more expensive than fossil-fuel furnaces).

The biggest operating concern in the use of channel furnaces is the rapid metal flow through the channel, which can lead to two opposing difficulties. The first is rapid wear of the refractory lining of the channel (Hellsing and Tallbäck, 1982;

Hentschel and Feldmann, 1982; MacIntosh, 1983), which causes the need for replacement at a more frequent rate than coreless units. The second is the deposition of dross on the refractory surface, similar to the deposition of dross on the crucible walls of coreless furnaces. This dross can build up over time and clog the channel (Whiteley, 2011), forcing the furnace to be taken out of service while the clog is removed. Solutions to the clogging problem include increasing the metal flow rate through the channel, which reduces dross deposition, and injecting an inert gas (argon or nitrogen) into the channel (Prillhofer, 2011). Both of these tactics increase the refractory wear rate, however. The problem is less severe in larger furnaces, which have wider channels and are less prone to clogging as a result.

RESISTANCE FURNACES

Electric resistance furnaces feature a wire element looped throughout the sides of the furnace. When current is passed through the wire, its resistance generates heat, raising its temperature. Eventually, the element temperature rises to the point where it glows. The glowing element radiates heat to the surroundings, including the crucible placed inside the furnace. This heats the metal inside or keeps it warm if already molten. Changing the current to the element changes the power generated, allowing temperature control.

The element in most resistance furnaces is made of the nickel–chromium alloy nichrome. The chromium in this alloy forms an impervious oxide layer when the element is used. The oxide layer prevents further oxidation and allows the element to be used for an extended period of time before replacement is required. The element is often installed as an *open-coil* element supported on high-purity ceramic tubes; elements can also be obtained semi-embedded in refractory panels (Atkins, 1992), which are simply swapped out when the element fails. Open-coil elements are preferred for aluminum melting, due to the higher power loading (and increased melt rate) possible with this type; however, improved semi-embedded elements are available that may eliminate this advantage.

Resistance furnaces generally belong to one of two categories: small-crucible *bale-out* furnaces and radiant-heat *electric reverberatory* holding furnaces and tundishes (Atkins, 1989; Butler, 2006; Morganite Crucible Ltd., 2006; Nabertherm, 2012; Schaefer, 2008). The capacity of the bale-out furnaces can be as high at 1500 kg but is usually much smaller. They can be operated with a molten heel but are usually charged cold. Efficiencies are high (80%), but the low power input means that melting capacity is poor compared to other furnaces. Because no stirring is provided by these furnaces, only bulky scrap can be used (along with primary ingot) in furnaces of this type. The lack of stirring also increases temperature gradients in the metal, which increases melt loss compared with induction furnaces (Puga et al., 2009). Crucibles are usually made from silicon carbide, which is preferred over clay–graphite because of its higher thermal conductivity.

Electric reverberatory furnaces (see Figure 10.6) are used as holding furnaces (Atkins, 1989; Butler, 2006; Morganite Crucible Ltd., 2006; Schaefer, 2008). They typically feature silicon carbide elements embedded into the furnace roof and can hold up to

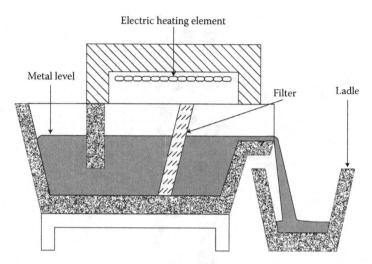

FIGURE 10.6 Cross section of a resistance holding furnace. (From Atkins, R., *Foundry Int.*, 15(4), 166, 1992. With permission.)

6000 kg of molten metal, although they are usually much smaller. The high reflectivity of molten aluminum means that a large bath surface is required for effective heating, and this in turn increases the potential for dross formation and hydrogen pickup during holding. As a result, use of these furnaces is limited.

Attempts have been made to overcome the disadvantages of resistance-heated furnaces. One of the more promising is a hybrid furnace that uses fossil-fuel heating for charge heating and melting and resistance heating for heating and holding the molten metal (Kanthal, 2009). This improves melting capacity, while preserving the electric heating advantages of cleaner metal and reduced gas pickup. A second is the *direct heat* concept (Chubu Electric Power Co., 2011; Free Library, 2012), which uses an induction coil to heat a silicon carbide crucible. The high resistivity of the crucible causes it to heat rapidly, heating the metal inside. Because the induction field is applied to the crucible rather than the metal, the stirring action of other coreless induction furnaces is avoided, reducing drossing and gas pickup.

Figure 10.7 shows a third alternative, developed over the past decade by an American consortium (Eckert, 2004). The figure shows a high-flux resistance heater mounted inside a ceramic tube. The resistance heater features a newly developed resistive material that has a higher melting point and better flexibility than previous options and a glass amalgam to provide better heat transfer. This heater is immersed into molten aluminum, where it transfers heat by conduction and convection rather than by radiation. Figure 10.8 shows a melting furnace arrangement that takes advantage of this technology. The furnace is the electric equivalent of a sidewell reverberatory furnace, with molten aluminum superheated by an array of immersion heaters being used to melt solid feed input to the charge well on the left (Osborne, 2010). However, the use of electric furnaces for large-scale melting of scrap is likely to remain the less preferred option for some time to come.

FIGURE 10.7 Immersion heater for isothermal melting process. (From Eckert, C.E., Aluminum success story: The isothermal melting process, March, 2004, http://www.apogeetechinc.com/wp-content/uploads/2010/09/IsothermalMelting.pdf. With permission.)

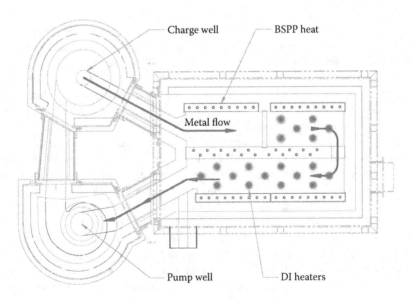

FIGURE 10.8 Isothermal melting process furnace architecture. (From Osborne, M., Energy efficient metal melting, transportation and holding in the casting plant, presented at *CastExpo '10*, Orlando, FL, March 22, 2010. With permission.)

RECOMMENDED READING

Atkins, R., Resistance heated melting and holding furnaces for aluminium casting, *Foundryman*, 82(2), 50, February 1989.

Butler, W.A., Melting and holding furnaces for die casting, *Die Cast. Eng.*, 50(3), 28, March 2006.

Perkul, R.Y., Induction furnaces, in *Metals Handbook, Vol. 15: Casting*, 10th edn., ASM International, Materials Park, OH, 2008, p. 108.

REFERENCES

Anderson, R.J., *Melting and Casting Aluminum*, Lindsay Publications, Inc. (reprint of 1925 original), Bradley, IL, 1987, p. 427.

Atkins, R., Resistance heated melting and holding furnaces for aluminium casting, *Foundryman*, 82(2), 50, February 1989.

Atkins, R., New developments in electric-resistance-heated melting and holding furnaces, *Foundry Int.*, 15(4), 166, April 1992.

Bala, K.C., Design analysis of an electric induction furnace for melting aluminum scrap, *AU Journal*, 9, 83, 2005.

Butler, W.A., Melting and holding furnaces for die casting, *Die Cast. Eng.*, 50(3), 28, March 2006.

Chubu Electric Power Co., Overview of development of IH aluminum melting and holding furnace, December 16, 2011. http://www.chuden.co.jp/english/corporate/ecor_releases/erel_pressreleases/__icsFiles/afieldfile/2011/12/16/0328.pdf.

Eckert, C.E., Aluminum success story: The isothermal melting process, March 2004. http://www.apogeetechinc.com/wp-content/uploads/2010/09/IsothermalMelting.pdf.

Fishman, O.S., Direct electric heat melting furnaces for aluminum and other non-ferrous metals, *Aluminium World*, 2(2), 41, February 2002.

Free Library, Aluminum direct heat electric melting furnace, October 30, 2012. http://www.thefreelibrary.com/Aluminum+Direct+Electric+Heat+Melting+Furnace.-a077829292.

Goyal, V., Aluminium melting with coreless medium-frequency induction furnaces, *Die Cast. Eng.*, 29(5), 32, May 1985.

Groteke, D.E., Aluminum crucible melt shop considerations, *Modern Cast.*, 87(12), 47, December 1997.

Heine, H.G. and Gorss, J.B., Coreless induction melting of aluminum, *Light Metal Age*, 49(1/2), 18, January/February 1991.

Hellsing, S. and Tallbäck, G., Compact high-power single-channel inductor for non-clogging operation on an aluminium furnace, in *Light Metals 1982*, Andersen, J.E., Ed., TMS-AIME, Warrendale, PA, 1982, p. 803.

Hentschel, F. and Feldmann, F., Application of the channel induction furnace for melting aluminum, *JOM*, 34(7), 58, July 1982.

Hipple, W.F., Melting with a medium frequency furnace, *Modern Cast.*, 72(8), 39, August 1982.

Kanthal, New highly efficient melting furnaces supplied to Die Casting Company, February 13, 2009. http://www2.kanthal.com/C12570A7004E2D46/062CC3B124D69A8EC1256988002A3D76/DDE6A857C0FA1F78C12570C20051578C/$file/6-D-4-3%20tubes%20stg.pdf?OpenElement.

Knödler, G., Induction furnaces for the melting of light metal, *Cast. Plant + Technol.*, 2(1), 22, January 1986.

Kreysa, E., Induction melting furnaces for aluminum, *Light Metal Age*, 49(7/8), 15, July/August 1991.

Lessiter, M.J., Aluminum crucible melting: Small foundry 'models', *Modern Cast.*, 87(12), 50, December 1997.

MacIntosh, W.H., Induction furnaces for melting secondary aluminium, *Conserv. Recycl.*, 6, 41, 1983.

Morganite Crucible Ltd., Morgan furnace range, August 11, 2006. http://refractoryandceramic.com.au/Morgan-Furnace-Range.pdf.

Nabertherm, Foundry, June 13, 2012. http://www.nabertherm.com/produkte/giesserei/giesserei_english.pdf.

Nealon, R.J., Caution: When melting aluminum (your energy savings deposit may be returned for inefficient funds), *Die Cast. Eng.*, 55(1), 37, January 2011.

Osborne, M., Energy efficient metal melting, transportation and holding in the casting plant, presented at *CastExpo '10*, Orlando, FL, March 22, 2010.

Perkul, R.Y., Induction furnaces, in *Metals Handbook, Vol. 15: Casting*, 10th edn., ASM International, Materials Park, OH, 2008, p. 108.

Prillhofer, B., Melt treatment of aluminium alloys in channel induction furnaces, February 20, 2011. http://www.amag.at/fileadmin/AMAG/AMAG/Pictures_NEW/AluReport/AR_2_09_EN_technology.pdf.

Puga, H., Barbosa, J., Soares, D., Silva, F., and Ribeiro, S., Recycling of aluminium swarf by direct incorporation in aluminium melts, *J. Mater. Process. Technol.*, 209, 5195, 2009.

Robiette, A.G.E., *Electric Melting Practice*, Charles Griffin & Co., London, U.K., 1972, Chapters 6–8.

Rowan, H.M., A new frequency selection chart for coreless induction furnaces, *Foundry Manag. Technol.*, 115(4), 34, April 1987.

Schaefer Furnaces, Furnaces for melting and holding aluminum, February 25, 2008. http://www.fwschaefert.com/document/CompleteBrochure2008.pdf.

Schifo, J.F. and Radia, J.T., *Theoretical/Best Practice Energy Use in Metalcasting Operations*, U.S. Department of Energy, Washington, DC, 2004. http://www1.eere.energy.gov/industry/metalcasting/PDFs/doebestpractice_052804.PDF.

Shuichi, B., Actual use of high frequency induction furnace, in *Light Metals 1988*, Boxall, L.G., Ed., TMS–AIME, Warrendale, PA, 1988, p. 413.

Smith, L. and Hayes, P.J., Recycling aluminium scrap in coreless induction furnaces, in *International Conference of Recycling Metals*, ASM International, Materials Park, OH, 1992, p. 151.

Whiteley, P., A historical perspective of aluminium casthouse furnace developments, *Mater. Sci. Forum*, 693, 73, 2011.

11 Recycling Industry

Some of the information in this chapter will be out of date before you read it. The recycling industry is constantly changing, and the number of facilities processing aluminum scrap or dross grows or shrinks with time in different locations due to several factors. As a result, some recyclers counted in this chapter are now defunct, and new ones have begun operations since this was written.

Secondary aluminum is a commodity material, and the driving forces that influence other commodity processors also influence the aluminum recycling industry. Most of these forces are rationalizing the industry, resulting in fewer but larger processing facilities. The changing nature of the scrap supply also affects which recycling facilities will thrive, as does the shifting geography of demand for aluminum.

WHO RECYCLES ALUMINUM?

As mentioned in Chapter 1, much of the aluminum scrap recycled is actually consumed by primary aluminum smelters rather than by scrap processors (Scharf–Bergmann, 2008). Primary smelters use scrap because of the following reasons:

- It is a cheap source of aluminum. In countries with high power costs and distant bauxite resources, the cost of purchasing and remelting scrap may be less than that of producing aluminum from ore (Koscielski, 2010; Wilkinson, 2003). As a result, a primary smelter can lower its costs by remelting as much scrap as possible. (Of course, the entry of primary smelters into the market can drive scrap prices up, reducing this advantage.) In countries with lower power costs or nearby bauxite resources, this factor becomes less important.
- It is a cheap source of alloying elements. As Chapter 2 pointed out, most aluminum alloys contain one or more of five alloying elements: copper, magnesium, manganese, silicon, and zinc. As the prices of these metals increase (silicon has been particularly expensive at times), the alloy content of scrap is increasingly valuable (Gaustad et al., 2007).
- It reduces power usage and emissions associated with primary smelting. Primary aluminum smelters consume substantial amounts of electrical energy and are often the biggest electricity consumers in a region. As other consumers in the region increase the demand for power, reducing power use in smelters is a net *contribution* to the available supply. In addition, many industrial societies have committed themselves to reducing carbon dioxide emissions to the atmosphere. As Chapter 1 pointed out, the CO_2 emitted during the recycling of aluminum is 90% less than that created during primary production (Damgaard et al., 2009; Das, 2006).

However, because primary smelters produce primarily wrought alloys and are not equipped to do substantial amounts of demagging, they are selective about the scrap they buy. New scrap with a known composition is vital; uncoated material is also needed, since primary smelters have different environmental equipment than secondary smelters and may not be able to adequately handle the organic vapors generated by decomposing coatings. As a result, primary smelters in developed areas with large scrap markets are more likely to use scrap than smelters in less-developed regions.

Some primary smelters in North America and Europe have also begun to produce casting alloys such as A356.0 (Grunspan, 2004; Rombach, 2002). The low iron content of primary aluminum makes it easier for these smelters to produce low-iron casting alloys, and the presence of nearby customers such as auto producers provides a further economic advantage. The production of cast rather than wrought alloys also means greater consumption of scrap, since these compositions can absorb more in the charge.

The second category of recycling operation is the *remelter*. Like primary smelters, remelters also produce mostly wrought alloys (Kevorkijan, 2003; Kirchner, 2002), meaning that careful selection of scrap grades and chemistries is essential. Again, remelters use mostly new scrap, with some added primary metal to dilute impurity content to the needed level (Das, 2006). Remelting operations fall into three categories:

1. *Integrated* operations generate a consumer product, often extrusions, rather than merely producing billets. Scrap generated during extruding is recycled along with purchased scrap.
2. *Nonintegrated* operations produce alloy billet for sale to other processors (Fielding, 2006).
3. *Toll* operators accept scrap from a single source and melt it down for return to the originator without ever actually owning the material. This type of operation has become more popular in recent years, as large manufacturers arrange with secondary aluminum producers to operate dedicated recycling facilities designed to process only the manufacturer's scrap (Carey, 2013; Petersen and McDonnell, 2008). About 10% of US secondary aluminum production comes from toll processors (Borner, 2012).

Initially, remelters were distinguished from secondary smelters by a lack of refining capability. However, the demand for higher product quality now means that many remelters now refine and filter their molten metal, regardless of its initial source.

A special type of remelter is the *UBC recycler,* which uses a feedstock consisting primarily of recovered beverage cans (Thornton et al., 2007). These first began to appear in the 1980s, when the number of discarded cans became sufficient to justify a specialized facility. UBC recyclers are similar to remelters in that they produce a wrought alloy, typically 3004. However, because their feedstock is a mixture of can bodies (3004) and lids (5175 and others), additions of primary aluminum and alloying elements are needed to produce the correct alloy composition. In addition, refining and filtration are required to produce a quality product. UBC recyclers are found mostly in Europe and North America, due to higher per capita use of cans there.

Secondary smelters, known as *refiners* in Europe, are the most common type of aluminum recycling facility. Secondary smelters produce casting alloys (Kirchner, 2002), in particular one of the 380 variants (see Chapter 4), and as a result use types of obsolete or low-grade scrap that other recycling facilities cannot use. As the demand for better-quality scrap from other recyclers increases its price, secondary smelters increasingly rely on old scrap to meet their needs (Buchanan, 2002). However, the use of old scrap makes quality control more challenging, and so refining and filtration are increasingly necessary. The use of old scrap also makes operating in an environmentally acceptable manner more difficult, since old scrap has more painted and coated material and is also more likely to contain plastics and salts.

INFLUENCES ON THE ALUMINUM RECYCLING INDUSTRY

IMPACT OF GOVERNMENT

Government decisions have a significant impact on where aluminum recycling facilities are located and how they do business. A few examples are listed as follows:

- The control of metal flow across a country's border is a common means of manipulating scrap-based industries (Alipchenko, 2009; Anon., 2007; De Oliveira, 2012; Kirchner, 2007b). The most common means of doing this is to restrict scrap exports, either with an export tax (Belarus, Brazil, China, Georgia, India, Russia, Pakistan, Vietnam, Zambia) or an outright ban (Azerbaijan, Ghana, Indonesia, Iran, Kazakhstan, Nigeria, Sri Lanka, Turkmenistan, Ukraine, Venezuela, Zimbabwe). This keeps scrap from leaving the country, making it available for local recyclers to use. The move to impose export controls has gained momentum in recent years, fueled by the aggressive purchasing of scrap by Asian recyclers. Export controls have other consequences as well. They lower the value of scrap, reducing the incentive to collect it, and they discourage the rationalization of the industry to larger and more efficient smelting or remelting facilities. Import taxes on refined aluminum also support local producers and have been in place for some time in the European Union (Pawlek, 2006).
- In 2000, the European Union enacted directive 2000/53/EC, better known as the End-of-Life Vehicles (ELV) directive (Kitagawa, 2011). This directive requires automotive manufacturers to take ELV vehicles back when their working life is finished and to ensure appropriate disposal (Gesing, 2005). The directive requires recovery or reuse of 85% (eventually 95%) of the material in the vehicle, with emphasis on reuse rather than reprocessing of materials. Similar requirements have been put in place in Japan. Since automakers will bear the cost of processing or disposal, recovery costs are important, and this encourages the development of closed-loop recycling, using a few favored scrap processors. This further reduces the number of small recycling operations in western Europe. Another possible response

by automakers is to sell ELVs to countries outside the European Union, ultimately encouraging recycling of their aluminum content elsewhere.

- On July 22, 2011, the US Environmental Protection Agency (EPA) proposed a rule that would legally redefine metal scrap as a solid waste material and equate recycling of scrap metal with that of hazardous waste (Spoden, 2013). Recyclers in both the United States and Europe have been struggling for years to convince government agencies that scrap is not waste and should not be regulated as such (Taylor, 2011b). The difference to metal recyclers can be significant. The definition of metal scrap as a raw material rather than waste in Europe exempts it from export prohibitions under the Basel convention (Muchová and Eder, 2010). On the other hand, the imposition of new regulatory requirements is likely to reduce development of new recycling programs. Because the economic advantage of using recycled metal over primary is small, measures that increase the cost of recycling often make it unviable, ultimately favoring the use of primary metal instead.

Other government mandates that impact the aluminum recycling industry include initiatives to increase recovery of UBCs, discussed in Chapter 3; required recycling of computers and other electronic devices; and initiatives to increase materials recovery from construction and demolition debris. These tend to support recycling facilities in industrialized countries, where these material streams are generated.

- In 2013, the Emissions Trading Scheme (ETS) created by the European Union will apply to the secondary aluminum industry as well as the primary industry (Luo, 2009). The ETS imposes charges on companies for the amounts of greenhouse gases (GHGs) their processes release, beyond a set *free allocation*. This free allocation will decrease with time and is scheduled to disappear entirely by 2027. Producers whose GHG emissions are less than that allowed can sell their unused allotment to those who have excess emissions, a system known as *cap-and-trade*; a market price exists for these emissions, and the incomes to be earned from selling them encourages even low GHG producers to continue to find ways to lower emission further. Similar schemes are under consideration by Australia, Canada, and the United States (Anton, 2010).

In general, ETS benefits the aluminum recycling industry. Secondary aluminum production produces roughly 90% less GHG than primary production (Damgaard et al., 2009; Kim et al., 2010), which allows primary producers to lower their overall impact and enhances the value of recycled metal. However, application of ETS to the secondary aluminum industry also has costs, especially for facilities that are less energy efficient (Luo, 2009). The application of GHG costs to typical secondary smelters could raise the total conversion cost by 5%–6%; in an environment such as Europe where recyclers are already under stress, the results could be continued reduction of the industry and movement of recycled aluminum production to countries where such additional costs are not incurred (Gesing, 2005). This would in turn result in increased import of aluminum to the EU from elsewhere and greater global

GHG emissions in the process. EU efforts to improve the sustainability of manufacturing industries create a similar risk of unintended consequences (Glimm, 2009).

IMPACT OF DEMAND

Secondary aluminum smelters exist primarily to serve the needs of the transportation industry, in particular auto producers; 70%–80% of their production is sold to this sector (De Oliveira, 2012). As a result, secondary smelters tend to be located near foundries serving the auto industry, and when the auto industry moves, the secondary smelting industry moves with it (Koscielski, 2010). This is especially apparent in eastern Europe and Mexico (Buchanan, 2002; Wilkinson, 2001). Figure 11.1 shows the location of major (>10,000 tpy) aluminum recycling facilities in India (squares) and that of auto production facilities (open circles). Nearly all of the secondary smelters are located near their primary customer, an automotive producer (Jhunjhunwala, 2009).

For remelters, the development of closed-loop recycling arrangements with scrap generators has meant more stability. This has meant less movement of this industry than for secondary smelters. UBC recyclers in particular have remained primarily in North America, since that is where both the scrap cans and the can producers are located. However, broader changes in metal demand affect this branch of the industry as well. The rapid and sustained growth of the Chinese economy in recent years has increased demand for all metals, and this has contributed to the sustained

FIGURE 11.1 Location of major aluminum recycling facilities (filled squares) and automotive manufacturers (open circles) in India.

International recycling activities and
scrap flows 2011

3,500,000 t*

*Estimate without remelted ingots

FIGURE 11.2   Global flow of aluminum scrap in 2011. (Courtesy of Günter Kirchner.)

increase in scrap flow to China (Gesing, 2005; Kirchner, 2007a,b). Figure 11.2 shows the global movement of aluminum scrap in 2011; the importance of demand from China is the predominant feature of global trade.

IMPACT OF COST

For aluminum recyclers, the cost of operation can be separated into two categories: the cost of obtaining scrap and the cost of processing it. As might be suspected, recyclers tend to be located in areas where these costs are less.

Some of the factors affecting the price of scrap have already been mentioned (see Chapter 4), including export or import fees and competition for scrap from primary smelters. Scrap prices are also affected by the price of primary aluminum and by the quality. (New scrap sells for more than old scrap, and sorted scrap sells for more than mixed scrap.) Sudden increases in scrap availability without an increase in nearby smelting or remelting capacity lower prices, encouraging recycling operations to set up shop. The increase in smelting capacity in eastern Europe has been accredited to this. As demand for scrap increases, the price eventually increases too, erasing this cost advantage.

The cost of operation is affected by the costs of energy, capital equipment, labor, and waste disposal. Since most secondary aluminum remelters and smelters fire their furnaces with natural gas, increases in natural gas price are important. Areas with lower energy prices become attractive, even if other costs are higher there. As pointed out in Chapter 6, lower labor costs in China made hand sorting more profitable than automated sorting in Europe and North America (Gesing, 2005; Schwalbe, 2011), although this is now changing; this is another reason why scrap supplies have been moving to China (Pawlek, 2006). Waste disposal costs are also higher in western Europe and North America, particularly for salt slags (Schwalbe, 1998), and this too encourages the industry to move elsewhere.

IMPACT OF TECHNOLOGY

The story of aluminum recycling has been largely the story of facilities attempting to manufacture a quality product suitable to customers. The most important techno-logical development has been the installation of refining and filtration equipment, which has made the quality of secondary aluminum close to that of primary. Another important technological development has been the construction of crucibles in which smelting and remelting facilities can send molten metal directly to customers (Hodge, 2009; Kuom and Urbach, 2007), without the need to cast it into ingots or billets. Shipping molten metal eliminates the expense of casting it and saves the customer money by eliminating the energy cost otherwise required to remelt it. However, a recycler shipping molten metal needs to be physically close to his customer, and this has provided further encouragement to recyclers to relocate as needed. Closed-loop recyclers are particularly subject to this restriction.

The most noticeable impact of improved technology can be seen in Table 11.1 (Scharf–Bergmann, 2007). Small *garage* secondary smelters used to be quite com-mon in Europe, collecting scrap over a small area and producing recycled secondary ingot (RSI) for sale to foundries. However, these small operations cannot afford to install the refining and environmental equipment increasingly needed to produce a quality product and meet government regulations. As a result, these operations have been falling by the wayside (Kirchner, 2006; Lokshin and Makarov, 2007; Scharf–Bergmann, 2008), while the number of large smelting facilities has been increasing.

RECYCLING AROUND THE WORLD

Figure 11.3 shows the location of *major* scrap aluminum consumers (>10,000 tons/year) on the African continent, from the listing by Pawlek (2012). Consumers include refiners, secondary smelters, and primary smelters using scrap, as well as foundries that consume large quantities of scrap directly. As can be seen, the industry is largely nonexistent in Africa, the result of a lack of supply of scrap, the lack of manufactur-ers to consume it, and a primary aluminum industry that can more than meet the continent's needs. There are two exceptions. The first is South Africa, where a small

TABLE 11.1
Secondary Smelter Numbers in Europe, 1994–2005

Year	Smelters with Capacity <5000 Tons/Year	Smelters with Capacity >20,000 Tons/Year
1994	162	23
2000	100	30
2005	48	31

Source: Scharf–Bergmann, R., Chances and risks of the European aluminium industry on the global market, presented at *9th OEA International Aluminium Recycling Congress*, Cologne, Germany, 2007. With permission.

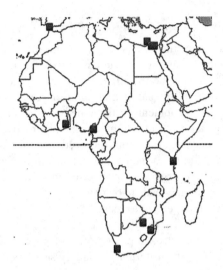

FIGURE 11.3 Location of major (>10,000 tpy) scrap aluminum consumers in Africa. (From Pawlek, R.P., *Light Met. Age*, June 26, 2012, http://www.lightmetalage.com/producers.php. With permission.)

collection of secondary smelters is struggling against higher scrap prices and loss of industrial capacity (Odendaal, 2010). The other is Egypt, where scrap is consumed mostly by foundries and some small secondary smelters (SEAM, 2005).

Figure 11.4 shows the major scrap consumers of South America. Most of the continent's capacity is concentrated in two countries, Argentina and Brazil. As is the case with Africa, South America is a net scrap exporter (De Oliveira, 2012), mostly to Asia. Most of the scrap consumed consists of UBCs, which Brazil does a better job of collecting than most other countries (Minter, 2010; Scharf–Bergmann, 2008).

FIGURE 11.4 Location of major scrap aluminum consumers in South America. (From Pawlek, R.P., *Light Met. Age*, June 26, 2012, http://www.lightmetalage.com/producers.php. With permission.)

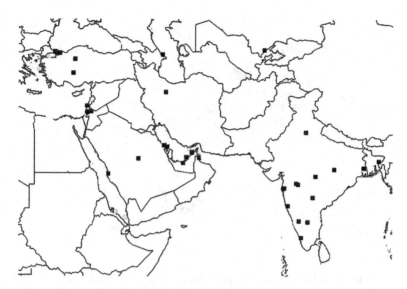

FIGURE 11.5 Location of major scrap aluminum consumers in south Asia and the Middle East. (From Pawlek, R.P., *Light Met. Age*, June 26, 2012, http://www.lightmetalage.com/producers.php. With permission.)

The UBC is not recycled directly but used by secondary smelters to produce casting alloy for the growing automotive industry.

The recycling industry of the Middle East and south Asia is highlighted in Figure 11.5. The industry is concentrated in three areas: India, Turkey, and the Gulf States. The structure of the Indian industry has been previously discussed; about 70% of recycled metal in India is cast alloy in the automotive sector (Gopalkrishnan, 2012). Labor costs are low in India and barriers to entry are few, so the industry features many small producers and is a scrap importer (Minter, 2006). Gulf area recyclers also import most of their scrap from the surrounding region (Minter, 2007; Scharf–Bergmann, 2007). Most of the scrap purchased in the Gulf is used by primary producers, both for mixing and for the separate production of cast alloys.

Figure 11.6 shows the recycling industry of Southeast Asia. As Hayashi (2012) points out, much of the growth in secondary aluminum production in the region is directly related to the growth of the auto industry. In particular, this has encouraged the growth of the industry in Malaysia (Minter, 2011a), which is now a net importer of aluminum scrap (Michida, 2011). Countries with smaller recycling industries struggle to keep scrap from being sold overseas, particularly to China (Minter, 2008a).

Figure 11.7 shows the locations of aluminum recyclers in east Asia. Expensive electrical power and the lack of domestic bauxite mean that almost all of Japan's domestic aluminum production consists of recycled metal. Hayashi (2012) states that about 20% of the scrap processed in Japan is imported, especially from Russia. The rest is obtained domestically. As is the case elsewhere, secondary aluminum production in Japan is tied to the auto industry, and as Japanese automakers increasingly locate their manufacturing facilities outside Japan, the aluminum recycling industry

FIGURE 11.6 Location of major scrap aluminum consumers in Southeast Asia. (From Pawlek, R.P., *Light Met. Age*, June 26, 2012, http://www.lightmetalage.com/producers.php. With permission.)

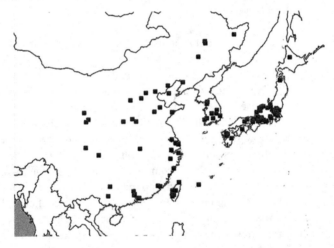

FIGURE 11.7 Location of major scrap aluminum consumers in east Asia. (From Pawlek, R.P., *Light Met. Age*, June 26, 2012, http://www.lightmetalage.com/producers.php. With permission.)

moves with them. Lower-grade scrap is often sent to China (Michida, 2011; Yoshida et al., 2005).

The South Korean recycled aluminum industry is closely tied to the Hyundai–Kia automotive group (Hayashi, 2012). Much of the recycled metal is shipped as molten casting alloy. Novelis recently began operations at a UBC direct recycling facility at Yeongju, the largest such facility in Asia (Watanabe, 2012).

As is the case with many other industries, the secondary aluminum industry in China has expanded dramatically since 2001 (Hayashi, 2012). An increasing fraction of the scrap used comes from domestic sources (Minter, 2008b), but the Chinese need for raw material has led them to become the most significant purchaser of scrap in several markets around the world, leading to the loss of aluminum recycling in several countries (Kirchner, 2007a,b; Scharf–Bergmann, 2007). The advantage of cheap Chinese labor for sorting and melting operations has been well noted; however, labor costs have risen significantly in the industry since 2010 (Minter, 2011b), and the need to meet previously ignored environmental standards evens the playing field still further. As a result, the Chinese secondary aluminum industry is beginning to move down a familiar path—closing smaller and dirtier operations, moving to more automated sorting technology, and improving product quality (Lili, 2008).

Figure 11.8 shows the location of major aluminum recyclers in Europe (Pawlek, 2012). The concentration of remelters and secondary smelters across the continent reflects the greater scrap supply from industrialized societies, as well as the decline of primary smelting as a source of aluminum (Conserva, 2011). Most secondary production comes from France, Germany, Italy, and the United Kingdom. The European secondary industry has suffered in recent years from loss of scrap to Asia, denial of scrap from eastern Europe by export restrictions, and high energy and labor costs. The presence of the auto industry is still important in determining the location of the industry (Kirchner, 2006); Lepeň (2011) has shown how most of the recycling capacity of eastern Europe is located within 200 km of eastern Czechia, the location of several automotive assembly plants. As previously mentioned, an increasing amount of scrap in Europe is consumed by primary producers, who can lower costs and meet environmental goals by doing so (Scharf–Bergmann, 2007).

Figure 11.9 shows the location of shredding facilities in North America (Taylor, 2011a). Their location corresponds to population density, which is proportional to the supply of shreddable goods. The shredding business is highly competitive, and it is thought that excess capacity may be shutting down over the next few years (Taylor, 2010). Figure 11.10 shows North American scrap consumers (Pawlek, 2012).

FIGURE 11.8 Location of major scrap aluminum consumers in Europe. (From Pawlek, R.P., *Light Met. Age*, June 26, 2012, http://www.lightmetalage.com/producers.php. With permission.)

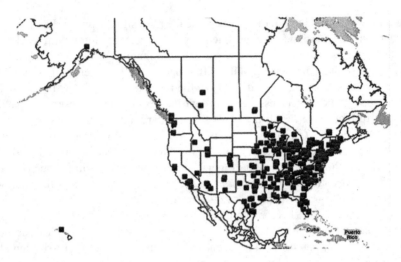

FIGURE 11.9   Location of scrap shredding facilities in the United States and Canada. (From Taylor, B., *Recycl. Today*, July 27, 2011a, http://www.recyclingtoday.com/FileUploads/file/RT-Auto-Shredder-Map-2010.pdf. With permission.)

FIGURE 11.10   Location of major scrap aluminum consumers in North America. (From Pawlek, R.P., *Light Met. Age*, June 26, 2012, http://www.lightmetalage.com/producers.php. With permission.)

The geographic concentration is somewhat different from that of the shredders, as it tends to follow the location of customers, particularly the auto industry (Garen et al., 2009). The closure of primary smelters in the United States has encouraged growth of the industry (Koscielski, 2010); 56% of US aluminum production in 2008 was secondary. The United States exports large amounts of scrap to Asia, in particular China; it also imports scrap from Canada and Mexico (Kirchner, 2007a), in particular UBC scrap.

RECOMMENDED READING

Gesing, A.J., Aluminum, recycling and transportation, presented at *Aluminium 2005*, Kliczow Castle, Poland, October 13, 2005. http://www.gesingconsultants.com/publications/50926.pdf.

Hayashi, S., Trends and future development of the secondary aluminium alloy industry in Asia, presented at *2012 BIR World Recycling Convention & Exhibition*, Rome, Italy, 2012. http://webmail.bir.org/birweb/assets/private/presentations/Rome2012/NonFerrousHayashi.pdf.

Kirchner, G., Aluminium recycling in Europe: Current situation and prospects, presented at *9th OEA International Aluminium Recycling Congress*, Cologne, Germany, 2007b. http://www.metalriciclo.com/Documenti/PDF_relazioni_convegni/Kirchner.pdf.

REFERENCES

Alipchenko, A., CIS aluminium recycling: Topical problems today and probable tomorrow, presented at *10th OEA International Aluminium Recycling Congress*, Berlin, Germany, February 3, 2009.

Anon., The global dimension of recycling becomes more and more important, *Aluminium*, 83(4), 64, 2007.

Anton, K., *Turning Carbon Exposure from Risk to Opportunity*, Alcoa Global Primary Products, January 28, 2010. http://utkstair.org/stair/presentations/anton_presentation.pdf.

Borner, G., Secondary scrap demand for aluminum, presented at *Platts Metals Week Aluminum Symposium*, Fort Lauderdale, FL, January 17, 2012. http://www.platts.com/IM.Platts.Content/ProductsServices/ConferenceandEvents/2012/gc203/presentations/Gary_Borner.pdf.

Buchanan, S., Mind the scrap gap! *MBM*, 379, 28–31, July 2002.

Carey, R., NASAAC—An analysis of a capitalist system gone wrong, *Die Cast. Eng.*, 57(1), 36–40, January 2013.

Conserva, M., The aluminium industrial system after the crisis, presented at *Aluminium 2000 7th World Congress*, Bologna, Italy, May 17, 2011.

Damgaard, A., Larsen, A.W., and Christensen, T.H., Recycling of metals: Accounting of greenhouse gases and global warming contributions, *Waste Manage. Res.*, 27, 773–780, 2009.

Das, S.K., Emerging trends in aluminum recycling: Reasons and responses, in *Light Metals 2006*, Galloway, T.J., Ed., TMS-AIME, Warrendale, PA, 2006, p. 911.

De Oliveira, R., Brazil: Leading the way in Latin American recycling, presented at *Metal Bulletin 20th International Recycled Aluminium Conference*, Salzburg, Austria, November 20, 2012.

Fielding, R.A.P., With MATALCO, Triple M defines the paradigm in aluminum recycling, *Light Met. Age*, 64(4), April 2006.

Garen, J., Jepsen, C., and Scott, F., *Economic Forces Shaping the Aluminum Industry*, Sloan Center for a Sustainable Aluminum Industry, Lexington, KY, July 2009. http://www.secat.net/sustainablealuminum.org/pdf/Economic_Forces.pdf.

Gaustad, G., Li, P., and Kirchain, R., Modeling methods for managing raw material compositional uncertainty in alloy production, *Res. Conserv. Recycl.*, 52, 180, 2007.

Gesing, A.J., Aluminum, recycling and transportation, presented at *Aluminium 2005*, Kliczow Castle, Poland, October 13, 2005. http://www.gesingconsultants.com/publications/50926.pdf.

Glimm, S., Recycled metal content—The Trojan horse for the sustainability of aluminium products, presented at *10th OEA International Aluminium Recycling Congress*, Berlin, Germany, February 3, 2009.

Gopalkrishnan, S., Strong trends in the Indian recycled aluminium business, presented at *Metal Bulletin 20th International Recycled Aluminium Conference*, Salzburg, Austria, November 20, 2012.

Grunspan, J.J., Brochot casting system for aluminum alloys, December 14, 2004. http://my.alacd.com/tms/2005/0PP1.pdf.

Hayashi, S., Trends and future development of the secondary aluminium alloy industry in Asia, presented at *2012 BIR World Recycling Convention & Exhibition*, Rome, Italy, 2012. http://webmail.bir.org/birweb/assets/private/presentations/Rome2012/NonFerrousHayashi.pdf.

Hodge, S., Aluminium flows, *MBM*, 458, 34, April 2009.

Jhunjhunwala, V., Aluminium recycling in India: Present scenario and future prospects, presented at *10th OEA International Aluminium Recycling Congress*, Berlin, Germany, February 3, 2009.

Kevorkijan, V., Recycling of wrought alloys, *Alluminio e Leghe*, 15(151), 61, July–August 2003.

Kim, H.-J., McMillan, C., Keoleian, G.A., and Skerlos, S.J., Greenhouse gas emissions payback for lightweighted vehicles using aluminum and high–strength steel, *J. Indus. Ecol.*, 14, 929, 2010.

Kirchner, G., The European and global dimension of aluminium recycling at present and in future, *Erzmetall*, 55, 465, 2002.

Kirchner, G., Aluminium recycling—A steady guarantee of aluminium supply, *Aluminium*, 82(12), 1194, 2006.

Kirchner, G., Aluminium recycling: The global dimension, presented at *9th OEA International Aluminium Recycling Congress*, Cologne, Germany, 2007a.

Kirchner, G., Aluminium recycling in Europe: Current situation and prospects, presented at *9th OEA International Aluminium Recycling Congress*, Cologne, Germany, 2007b. http://www.metalriciclo.com/Documenti/PDF_relazioni_convegni/Kirchner.pdf.

Kitagawa, K., *Automobile Recycling in Europe, Japan & North America*, Japan Productivity Center, October 9, 2011. http://www.ine.gob.mx/descargas/dgcenica/2011_09_09_kkitagawa.pdf.

Koscielski, M., Unwrought aluminum: Industry & trade summary, United States International Trade Commission, March 2010. http://www.usitc.gov/publications/332/ITS_6.pdf.

Kuom, M. and Urbach, R., Erweterte belieferungsmöglichkeiten von aluminiumgießereien mit flüssigaluminium, *Giesserei*, 94, 162, 2007.

Lepeň, L., Aluminium–recycling in Ost–und Zentraleuropa, presented at *11th OEA International Aluminium Recycling Congress*, Vienna, Austria, 2011.

Lili, S., With meager profits only the fittest will survive, *Aluminium*, 84(3), 50, March 2008.

Lokshin, M. and Makarov, G., Status and prospects for the development of aluminium recycling in Russia, presented at *Aluminium 2000 6th World Congress*, Florence, Italy, March 15, 2007.

Luo, Z., The impact of emissions trading on the secondary aluminium industry of the EU, *Aluminium*, 85(3), 21, March 2009.

Michida, E., International trade of recyclables in Asia: Is cross–border recycling sustainable? Chapter 2 in *Economic Integration and Recycling in Asia: An Interim Report*, Kojima, M. and Michida, E., Eds., IDE–JETRO, Chiba, Japan, 2011. http://www.ide.go.jp/Japanese/Publish/Download/Report/2010/pdf/2010_431_02.pdf.

Minter, A., India's scrap struggle, *Scrap*, 63(6), 46, November/December 2006.

Minter, A., Sultans of scrap, *Scrap*, 64(3), 82, May/June 2007.

Minter, A., Thailand's taste for scrap, *Scrap*, 65(5), 78, September/October 2008a.

Minter, A., Scrap made in China, *Scrap*, 65(2), 154, March/April 2008b.

Minter, A., Brazil rising, *Scrap*, 67(3), 58, May/June 2010.

Minter, A., Malaysia in the middle, *Scrap*, 68(3), 52, May/June 2011a.

Minter, A., The industrial revolution, *Scrap*, 68(2), 142, March/April 2011b.

Muchová, L. and Eder, P., *End–of–Waste Criteria for Aluminium and Aluminium Alloy Scrap: Technical Proposals*, JRC Scientific and Technical Reports, EUR 24396, 2010. http://ftp.jrc.es/EURdoc/JRC58527.pdf.

Odendaal, N., Call to impose an export tariff on scrap metal, *Engineering News Online*, September 2010. http://www.engineeringnews.co.za/print-version/the-decline-of-the-secondary-aluminium-industry-2010-09-24.

Pawlek, R.P., European secondary aluminium industry under threat, *Aluminium*, 82(1/2), 22, January/February 2006.

Pawlek, R.P., Secondary aluminum smelters of the world, *Light Met. Age*, June 26, 2012. http://www.lightmetalage.com/producers.php.

Petersen, T. and McDonnell, M., Beck, Met-Al ink scrap, secondary aluminum toll agreement, *Beck Aluminum Industry News Notes*, September 12, 2008. http://www.beckaluminum.com/news3.htm.

Rombach, G., Future availability of aluminum scrap, in *Light Metals 2002*, Schneider, W., Ed., TMS-AIME, Warrendale, PA, 2002, p. 1011.

Scharf–Bergmann, R., Chances and risks of the European aluminium industry on the global market, presented at *9th OEA International Aluminium Recycling Congress*, Cologne, Germany, 2007.

Scharf–Bergmann, R., Aluminium takes new directions, *MBM*, (447), 24, April 2008.

Schwalbe, M., The German secondary smelting business today and tomorrow: Meeting the competitive challenge, in *6th Secondary Aluminum Conference, Metal Bulletin*, 1998.

Schwalbe, M., The contribution of modern recycling technologies to increase recycling rates of aluminium, presented at *11th OEA International Aluminium Recycling Congress*, Vienna, Austria, February 21, 2011.

SEAM, *Report on the Secondary Aluminium Sector for Utensils in Egypt*, Egyptian Environmental Affairs Agency, Cairo, Egypt, February 14, 2005. http://ripecap.net/Uploads/455.pdf.

Spoden M.C., *EPA Solid Waste Proposal Will Discourage Recycling*, Stites & Harbison PLLC, February 12, 2013. http://www.environmentallawnews.com/files/2013/02/EPA-Solid-Waste-Proposal-Will-Discourage-Recycling.pdf.

Taylor, B., Moving target, *Recyl. Today*, April 20, 2010. http://www.recyclingtoday.com/Article.aspx?article_id=100465.

Taylor, B., Dispersal pattern, *Recyl. Today*, July 27, 2011a. http://www.recyclingtoday.com/FileUploads/file/RT-Auto-Shredder-Map-2010.pdf.

Taylor, B., War of words, *Recyl. Today*, November 21, 2011b. http://www.recyclingtoday.com/rt1111-waste-recycling-legislation.aspx.

Thornton, T., Hammond, C., van Linden, J., Campbell, P., and Vild, C., Improved UBC melting through advanced processing, in *Light Metals 2007*, Sørlie, M., Ed., TMS-AIME, Warrendale, PA, 2007, p. 1191.

Watanabe, M., Novelis starts up new South Korea aluminum recycling plant, *Platts News Analysis*, October 24, 2012. http://www.platts.com/RSSFeedDetailedNews/RSSFeed/Metals/7189518.

Wilkinson, J., Secondary aluminum: Despite structural changes, auto market remains key, *Am. Met. Market*, November 11, 2001.

Wilkinson, J., Sourcing far and wide, *MBM*, (395), 39, November 2003.

Yoshida, A., Terazono, A., Aramaki, T., and Hanaki, K., Secondary materials transfer from Japan to China: Destination analysis, *J. Mater. Cycles Waste Manage.*, 7, 8, 2005.

12 Metal Refining and Purification

As Chapter 4 pointed out, aluminum alloys produced by remelting scrap are less valuable than alloys with the same composition produced from primary metal. The reason for this is concern over the purity of the recycled metal alloy, which is often inferior to that of primary. The development of refining technology for molten aluminum is designed to eliminate this deficiency, which will allow recycled metal to compete with the primary in more applications. Much of the technology used for refining molten aluminum has been introduced only in the last 30 years, and what was once a minor footnote in an aluminum production flowsheet is now a major consideration.

The choice of refining strategy and technology used by scrap remelters depends on several factors—the type of scrap being remelted, the type of furnace used, the type of product being generated, and, most importantly, the needs of customers. As a result, there is no universal refining technology. A complicating factor is the blending of scrap with primary metal in melting furnaces, which also combines the refining concerns of the two types of metal. Because of this, much of the technology used for refining remelted aluminum scrap is also used to purify molten primary metal. Since the basic principles are the same, a discussion of refining technology for secondary aluminum is essentially a discussion of molten aluminum refining in general.

The following discussion is a brief introduction to the theory and equipment used for molten aluminum refining. Enough in-depth information exists on this technology alone to write an entire book about it. Those wishing to learn more about aluminum refining are encouraged to consult the references at the end of this chapter, in particular the review by Zhang et al. (2011).

COMMON IMPURITIES IN MOLTEN ALUMINUM

Table 12.1 lists the most common impurities in molten aluminum (Waite, 2002) and compares their concentration in primary and secondary metal. The impurities can be divided into three classes—hydrogen, reactive metals (including magnesium), and inclusions.

Hydrogen: As previously discussed, dissolved hydrogen in molten aluminum is obtained from a reaction between water vapor and the molten aluminum (Foseco, 2011; Fruehan and Anyalebechi, 2008):

$$3H_2O + 2Al = Al_2O_3 + 6\underline{H} \tag{12.1}$$

TABLE 12.1

Common Impurities in Primary and Secondary Molten Aluminum

Impurity		Concentration in Primary Metal	Concentration in Secondary Metal
Hydrogen		0.1–0.3 wppm	0.4–0.6 wppm
Inclusions (PoDFA scale)		>1 mm²/kg (Al_4C_3)	0.5 < mm²/kg < 5.0 (Al_2O_3, MgO, $MgAl_2O_4$, Al_4C_3, TiB_2)
Alkali	Sodium	30–150 ppm	< 10 ppm
	Calcium	2–5 ppm	5–40 ppm
	Lithium	0–20 ppm	< 1 ppm

Source: Waite, P., Technical perspective on molten aluminum processing, in *Light Metals 2002*, Schneider, W., Ed., TMS-AIME, Warrendale, PA, 2002.

The magnesium in remelted alloy scrap also reacts with water vapor:

$$H_2O + \underline{Mg} = MgO + 2\underline{H} \tag{12.2}$$

Both reactions are highly favored thermodynamically and limited only by the formation of an oxide skin on the melt surface, which prevents contact between the water vapor and the molten metal. Factors that encourage these reactions and increase the dissolved hydrogen content of the metal include the following:

- A higher vapor pressure of water vapor in the atmosphere (i.e., higher humidity), which drives the reactions to the right (Foseco, 2011)
- Metal turbulence, which destroys the oxide skin and allows the reactions to continue
- Wet or damp charge materials, which make water vapor directly available to the melt

In addition to natural humidity, water vapor in the products of combustion (POC) from fossil-fuel fired furnaces is a source of hydrogen (Enright, 2007; Foseco, 2011). As a result, extended exposure to POC in transfer ladles can raise dissolved hydrogen content still further. Some alloying elements (Cu, Fe, Si, Zn) raise the activity coefficient of dissolved hydrogen in molten aluminum, decreasing its solubility (Fruehan and Anyalebechi, 2008); others (Li, Mg, Zr) lower the activity coefficient and raise the solubility. Sigworth et al. (2008) have shown that 2000-series alloys (see Chapter 2) have a hydrogen solubility up to 30% lower than that of pure aluminum, and 5000 series can dissolve hydrogen at levels 50% or higher than the pure metal. As Table 12.1 shows, the hydrogen content of remelted scrap is usually higher than that of primary metal. This is caused mostly by water in the scrap and by the use of fossil fuels for remelting it (Pyrotek, 2006).

The unit for hydrogen analysis in molten aluminum is unusual. Hydrogen contents are often expressed as standard cm^3 of H$_2$ gas contained per 100 gm of molten aluminum (Fruehan and Anyalebechi, 2008). The reason is that hydrogen is virtually insoluble in solid aluminum. Because of this, any hydrogen dissolved in the molten metal will exsolve during solidification, resulting in porosity in the cast product (Enright, 2007). Since 100 g of solid aluminum has a theoretical volume of 37.04 cm^3, a dissolved hydrogen content of 0.5 cm^3/100 g would result in a porosity level of 1.33% in the solid product, unacceptable for many applications. Customer specifications often call for $\underline{H}$ levels in purchased aluminum of 0.18 cm^3/100 g (0.16 wppm) or less for wrought-alloy applications (Enright, 2007; Fielding, 1996), and this specification continues to decrease as rolled sections get thinner and less tolerant of porosity. Allowable hydrogen levels for casting alloys are higher (0.4 cm^3/100 g) but still lower than the level initially present in most molten scrap. Dissolved hydrogen levels in scrap melts can often exceed 0.5 cm^3/100 g (0.45 wppm).

The development of new analytical technology has changed the way aluminum producers and manufacturers analyze the hydrogen content of their metal. The goal has been to develop an instrument that can be used online, can provide results in real time, is inexpensive to use, can survive in a plant environment, and can provide reliable results. The analytical options resulting from this development (Apelian, 2009; Hills et al., 2009; Neff, 2004):

- The *reduced pressure test* (also known as the Straube–Pfeiffer test), which relies on Archimedes principle and assumes that all of the hydrogen in solution in the molten aluminum will come out as it solidifies. The reduced pressure test involves solidifying a small sample taken from the melt in a closed chamber under a set pressure (Enright, 2007; Velasco and Montalvo, 2012). Measuring the density of the solid determines its porosity; knowing the volume occupied by the pores and the pressure determines the amount of hydrogen released by the solidifying metal; knowing the mass of the aluminum allows calculation of the hydrogen concentration. However, in practice the reduced pressure test does not provide an accurate measurement of dissolved hydrogen content. Several factors determine the amount of porosity that forms in a solidifying sample, and at best the results are more useful for comparative purposes. In addition, the method does not provide real-time results. The reduced pressure test is popular in many foundries due to its simplicity and quick turnaround but may wind up being phased out as more accurate techniques become affordable.
- *Direct determination* methods measure the amount of H$_2$ generated by a molten sample. The best example of this is the Leco device. This device melts a solid sample taken from the furnace in a stream of helium. Hydrogen is transferred to the gas stream and passed over heated copper oxide, which reacts with the hydrogen to generate copper metal and water. Measuring the amount of water in the gas determines the amount of hydrogen in the sample. This produces a direct measurement of hydrogen content, rather than p_{H2}, and so does not require alloy- and temperature-dependent corrections. However, it does not operate online and does not produce quick

turnaround. In addition, the use of a solidified sample from the furnace raises the possibility that the hydrogen content of the sample and that of the melt are not the same.

- *Analytical* methods determine the partial pressure of H_2 exsolved from a sample into a vacuum or carrier gas. The best-known example of this is the Alscan device (Badowski and Droste, 2009; Enright, 2007; Exebio et al., 2008; Gansemer et al., 2007). This is an online measurement tool, in which a probe is inserted into the melt containing a ceramic foam. Dissolved hydrogen diffuses into the foam and is removed by nitrogen carrier gas. The hydrogen content of the equilibrated gas (and thus the partial pressure of hydrogen) is determined by measuring its thermal conductivity and comparing with a calibration chart. The process requires 10–15 min to complete, which limits its usefulness for process control (Hills et al., 2009). The accuracy of direct determination measurement is complicated by the need to know the relationship between p_{H_2} and dissolved hydrogen content as a function of alloy content and metal temperature.

- *Electrochemical* analysis is the newest category of hydrogen analysis method and has become more widely used since the previous edition of this book (Badowski and Droste, 2009; Hills et al., 2009; Pascual, 2009). The best-known example of this type of measurement is the Alspek device (Moores, 2008). This uses a solid electrolyte (indium-doped calcium zirconate) which conducts protons (H^+ ions) generated by the dissolved hydrogen in the molten metal; the higher the hydrogen content, the greater the ionic current generated in the electrolyte. An internal reference (a mixture of hydrogen-containing β-Zr and δ-ZrH$_2$) eliminates the need for an external hydrogen reference. This device can be used online, generates quick results, and is increasingly reliable. Its use will likely become more prominent over the next few years.

Reactive metals: Although this group of impurities in theory includes all the alkali and alkaline earth metals, in practice four elements matter most—sodium, calcium, lithium, and magnesium (Pyrotek, 2006). The sodium content of primary metal results from interaction between the aluminum generated in molten salt electrolytic cells and the sodium in the cryolite bath. Since this bath is not used to remelt scrap, secondary metal has little or no sodium. Some calcium is found in remelted scrap, however. The most significant source of calcium in scrap is the calcium-containing salts used for deicing roads and bridges, which can wind up attached to aluminum car and truck parts. Because primary metal is often added to molten scrap, the presence of sodium is often a concern as well. Lithium in primary metal is a decreasing concern, as the number of potlines using Li_2CO_3 continues to decrease. However, increasing use of lithium-containing alloys by the aircraft industry (see Chapter 2) means the eventual return of these alloys as scrap. Allowable lithium concentrations in automotive casting alloys are <3 ppm (Carey, 2013), so lithium removal is becoming more important. Sodium is deliberately added to a few aluminum alloys, but allowable sodium concentration in most others is 5 ppm or less (Dewan et al., 2011; Enright, 2007).

The amount of magnesium in molten scrap depends on the alloys being remelted and can range to 5% or higher. Whether this magnesium is treated as an impurity to be removed depends on the alloy being created; because magnesium is more expensive than aluminum, alloy practice that produces alloys with higher magnesium content than that of the melt is encouraged. However, in some cases, this is not feasible, and demagging is a part of the refining process.

As is the case with hydrogen, analysis of the reactive-metal content of molten aluminum is increasingly conducted on the shop floor, and quick turnaround time is important. The most popular technique is optical emission spectrometry (OES), which can also measure the concentration of other trace metals (Logan, 2012). OES measures relative concentrations of trace elements, so a good reference material is a must (Hamouche, 2004). The automation of the analytical process has been a significant improvement.

Inclusions: Inclusions are mostly solid particles suspended in molten aluminum. The number and size of these particles depends on a variety of things, in particular the initial quality of the scrap being melted and the impurities contained in that scrap. Inclusions are nonmetallic particles, usually less than 100 μm in size. They consist mostly of oxides, although several other types of compounds are represented. There are two basic classes of inclusion: *exogenous* and *indigenous*.

Exogenous inclusions are particles already existing as a separate phase before melting (Eckert, 1992). Small pieces of furnace refractory that break off into the melt are the best-known examples (Weaver, 1997); bits of oxide or dirt attached to scrap are another. Exogenous inclusions consist almost entirely of oxide and are much larger than most indigenous inclusions. Because of this, their presence in aluminum is more harmful than that of indigenous inclusions; however, their larger size makes them easier to remove.

Indigenous or *in situ* inclusions are formed by chemical reactions taking place in the melt (Eckert, 1992; de la Sabloniere and Samuel, 1996). An example is the reaction of dissolved oxygen with the molten aluminum to generate alumina:

$$2Al + 3\underline{O} = Al_2O_3 \tag{12.3}$$

The thermodynamic stability of alumina increases as the temperature decreases, and the solubility of oxygen in molten aluminum decreases with temperature as well, so reaction (12.3) automatically occurs in aluminum as it cools. In magnesium-containing alloys, magnesia and spinel are also created (Enright, 2007; Fielding, 1996):

$$\underline{Mg} + \underline{O} = MgO \tag{12.4}$$

$$\underline{Mg} + 2Al + 4\underline{O} = MgAl_2O_4 \tag{12.5}$$

If nitrogen is used during degassing, AlN can form; if $MgCl_2$ is used as a flux, chloride inclusions can form as well. A particular problem is TiB_2, formed when boron is alloyed with the metal for grain refining (Pyrotek, 2006). These inclusions are much smaller than exogenous inclusions and much harder to remove.

The inclusions in remelted aluminum scrap are different than those in primary metal, as Table 12.1 shows. Scrap inherently has more dirt and oxide in it than primary metal (Pyrotek, 2006), and the oxide skins generated during melting increase the inclusion count even more (Altenpohl, 1998). As a result, inclusion removal is much more important in secondary aluminum production. Remelting of bulky scrap generates metal with fewer inclusions than remelted UBC or light scrap; the price of bulky scrap is higher as a result (see Chapter 4). Remelting of high-magnesium alloys generates a higher inclusion count, due to the greater tendency of magnesium to react during melting.

Figure 12.1 describes in greater detail the significance of different types of inclusions in molten aluminum (Altenpohl, 1998). The hatched area is an unofficial *quality limit* for the concentration of inclusions of different sizes in cast aluminum. (The limits are lower in wrought aluminum.) The y-axes are inclusion concentration per unit area of as-cast microstructure; the x-axis is the average diameter of the inclusion. Curve *C* is for inclusions generated during casting, that is, oxide skins and slag inclusions; exogenous inclusions would also fall along this curve. Curve *B* is for indigenous inclusions generated during melting and melt treatment. In general, oxide inclusions tend to be larger than those composed of carbide or boride particles and the ones most in need of removal for quality standards.

The unit of inclusion analysis (mm²/kg) in Table 12.1 is unusual. The reason for this is the choice of analytical method. Before the 1980s, inclusions were counted by analyzing the microstructure of cast metal and by hand. However, the small number of inclusions made this a difficult and time-consuming process. In the 1980s, the introduction of *PoDFA* and *LAIS* made inclusion analysis easier (Enright, 2007; Neff, 2004; Poynton et al., 2009). PoDFA (Porus disk filtration apparatus) and LAIS (Liquid aluminum inclusion sampler) use the vacuum filtration of a known amount of

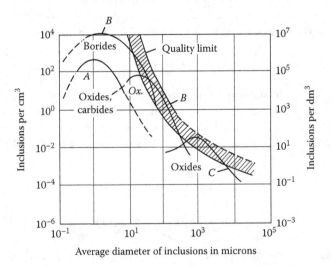

FIGURE 12.1 Concentrations of different types of inclusions in molten aluminum by particle size. (From Altenpohl, D.G., *Aluminum: Technology, Applications, and Environment*, 6th edn., TMS-AIME, Warrendale, PA, 1998. With permission.)

molten metal through a porous filter to collect the inclusions, which are then counted and measured. The total area of inclusion per kilogram of metal filtered is the variable measured and used in Table 12.1. These tests are inexpensive to conduct, but time consuming and expensive to analyze. As a result, their effectiveness as process management tools is limited.

A more useful tool is the Prefil Footprinter (Enright, 2007; Poynton et al., 2009), which also uses the forced passage of metal through a filter. In addition to collecting inclusions, Prefil measures the amount of metal passing through the filter as a function of time. A plot of metal flow rate against time indicates the relative number and size of inclusions. The shape of the initial portion of the curve is determined by the number of larger inclusions (Prillhofer et al., 2008); the number of smaller inclusions is apparent from the shape of the curve in the latter stages. Although Prefil does not give precise information about the number and size of inclusions, it is a useful tool for obtaining real-time results on the effect of process changes on the quality of molten aluminum at various stages in the melting and casting process.

A more recent development has revolutionized inclusion measurement in molten aluminum. *LiMCA* (Liquid metal cleanliness analyzer) is based on a Coulter counter, which detects inclusions by their impact on the electrical conductivity of aluminum flowing through a small tube (Engh, 1992; Poynton et al., 2009). Each inclusion is registered separately, allowing them to be counted per unit mass of metal. Within limits, LiMCA can also determine the size of inclusions and provide real-time analysis. Challenges include distinguishing between inclusions and gas bubbles and using LiMCA to detect and count inclusions smaller than 20 μm (Badowski and Instone, 2012). Considerable research has been performed on the development of ultrasonic sensors (Kurban et al., 2005), which would sample a larger liquid metal volume than LiMCA and provide quicker results; however, the high temperature of molten aluminum has made it difficult to develop a reliable instrument (Poynton et al., 2010).

The fourth class of impurities: Although not listed in Table 12.1, there is a fourth class of impurities that remelters would like to remove if possible. These are the alloying elements present in aluminum scrap—copper, iron, manganese, silicon, and zinc (Pyrotek, 2006). Percentages of these elements vary, depending on the scrap charge and the amount of primary metal added. However, concentrations of these elements that are too high limit the number of alloys that can be produced from a melt, as discussed in Chapter 2.

The development of technology for segregating scrap by alloy type was discussed in Chapter 5; however, the commercial use of this technology is in its infancy. In the meantime, the only current response to high concentrations of alloying elements is either the production of an alloy with even higher concentration of these elements, such as 364.1 or 380.1, or dilution of the alloying elements to a lower level with added primary metal.

Several efforts have been made in recent years to develop an approach for removing dissolved iron from molten aluminum (Zhang and Damoah, 2011). Gao et al. (2009) and Chen et al. (2012) demonstrated an electroslag refining process using sodium borate. The reaction between the borate and dissolved iron generates solid iron boride and sodium dissolved in the melt. A more common approach is to add manganese and silicon; this causes the precipitation of the intermetallic phase

$Al_{15}(Fe,Mn)_3Si_2$, which can be removed by filtration (Lee et al., 2010; de Moraes et al., 2006). While these techniques show promise for iron removal, they introduce new impurities that themselves have to be removed, and the added expense makes producing aluminum alloy from primary metal economically preferable. More improvement will be needed before these techniques become commercially viable.

FUNDAMENTALS OF IMPURITY REMOVAL

Hydrogen: The only successful technique for removing hydrogen from molten aluminum is to transfer it to a gas phase by the following reaction:

$$2\underline{H} = H_2 \text{ (g)} \tag{12.6}$$

The equilibrium constant for this reaction can be rearranged to a form of *Sievert's law*:

$$[\underline{H}] = S \, (p_{H_2})^{1/2} \tag{12.7}$$

where
  [H] is the dissolved hydrogen concentration in $cm^3/100$ g of aluminum
  p_{H_2} is the partial pressure of hydrogen in atm
  S is the equilibrium constant for Equation 12.6 at a given temperature

Sigworth et al. (2008) suggest that the most accurate expression of S in pure aluminum as a function of temperature is

$$\log S = \frac{-2632}{T} + 2.726 \tag{12.8}$$

where T is expressed in K. At 700°C (973 K), $S = 0.911$, and molten pure aluminum with a dissolved hydrogen concentration of 0.4 $cm^3/100$ g (0.36 wppm) will generate p_{H_2} equal to 0.156 atm. As long as the actual partial pressure of hydrogen is less than that, degassing will proceed; when the equilibrium level is reached, the overall reaction will stop. Producing a hydrogen level of 0.2 $cm^3/100$ g (0.18 wppm) requires a partial pressure of H_2 less than 0.048 atm.

The rate at which reaction (12.6) proceeds is a function of driving force (the difference between the equilibrium p_{H_2} and the actual p_{H_2} in the vapor phase), the temperature, and the relative surface area of interface between melt and vapor phase per unit volume of melt. Removal rates can be increased by

- Reducing the partial pressure of H_2 in the gas, either by using a vacuum or bubbling gas into the melt
- Raising the temperature (although this increases energy costs and raises $\underline{H}$ solubility)
- Generating higher relative surface area by adding lots of very tiny bubbles to the melt (Engh, 1992; Sigworth et al., 2008)

It was thought that the use of chlorine in degassing atmospheres might further enhance hydrogen removal (Peterson et al., 1995) through the following reaction:

$$2\underline{H} + Cl_2 = HCl \ (g) \tag{12.9}$$

However, HCl is thermodynamically less stable than $AlCl_3$, so the likelihood of chlorine being effective for hydrogen removal is doubtful. Several experiments have reached the conclusion that chlorine use in degassing, while useful for alkali removal, is no more effective at hydrogen removal than any other gas and might be detrimental (Chesonis and DeYoung, 2008; Williams et al., 2000).

Reactive metals: Figure 12.2 shows the logarithm of the equilibrium constant at 727°C for the reaction (Utigard, 1991)

$$Al + \left(\frac{3}{n}\right)MX_n = AlX_3 + \left(\frac{3}{n}\right)M \tag{12.10}$$

where
  M is another metal
  X is either chlorine or fluorine

The graph shows that aluminum is less reactive with chlorine and fluorine than the reactive metals contained in it (as evidenced by the negative "$\log K_{eq}$" values of the y-axis), and as a result added chloride or fluoride is more likely to react with

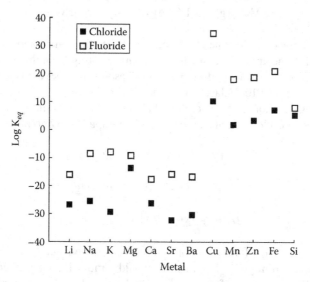

FIGURE 12.2 Logarithm of the equilibrium constant at 727°C for the reaction of molten aluminum with impurity–metal fluorides. (From Utigard, T., Thermodynamic considerations of aluminum refining and fluxing, in *Extraction, Refining, and Fabrication of Light Metals*, Sahoo, M. and Pinfold, P., Eds., CIM, Montreal, Canada, 1991. With permission.)

calcium, lithium, and sodium. This is why a molten sodium–aluminum fluoride salt bath is used for the electrowinning of primary aluminum and why lithium salts can be added to that bath. It also forms the basis for the removal of impurity calcium, lithium, and sodium from molten aluminum scrap. Chlorine gas is added and reacts with the impurities to form sodium or calcium chloride, which floats to the salt slag on top:

$$\underline{Ca} + Cl_2 = CaCl_2 \text{ (slag)} \qquad (12.11)$$

$$\underline{Na} + \tfrac{1}{2}Cl_2 = NaCl \text{ (slag)} \qquad (12.12)$$

Chlorine also reacts preferentially with magnesium and is the key to demagging of molten aluminum scrap as well:

$$\underline{Mg} + Cl_2 = MgCl_2 \text{ (slag)} \qquad (12.13)$$

However, the use of chlorine gas also presents environmental concerns, since it also reacts with the aluminum to generate aluminum chloride vapor:

$$2Al + 3Cl_2 = 2AlCl_3 \text{ (g)} \qquad (12.14)$$

This reaction occurs when the concentration of other impurities is low and so occurs in low-magnesium molten scrap or primary metal. The aluminum chloride reacts with water vapor above the melt to generate hydrochloric acid vapor (Eckert, 1992):

$$2AlCl_3 \text{ (g)} + 3H_2O \text{ (g)} = Al_2O_3 \text{ (s)} + 6HCl \text{ (g)} \qquad (12.15)$$

In addition, unreacted chlorine also winds up in the atmosphere above the melt. The workplace hygiene and environmental concerns resulting from this are an incentive to replace chlorine gas as a refining agent (Chesonis and DeYoung, 2008; Leboeuf et al., 2007; Ohno, 2010; Velasco and Nino, 2011).

Fluorine gas is not used for refining, but solid fluxes containing aluminum fluoride have been developed that can react with the impurities (Zholnin et al., 2005):

$$2AlF_3 + 3\underline{Ca} = 2Al + 3CaF_2 \text{ (slag)} \qquad (12.16)$$

$$AlF_3 + 3\underline{Na} = Al + 3NaF \text{ (slag)} \qquad (12.17)$$

$$2AlF_3 + 3\underline{Mg} = 2Al + 3MgF_2 \text{ (slag)} \qquad (12.18)$$

Because of this, AlF_3 fluxes cannot be used to remove calcium and sodium from magnesium-containing alloys—the AlF_3 would remove the magnesium as well. In addition, AlF_3 is expensive for secondary smelters, and the use of fluorides presents potential workplace hygiene concerns. As a result, fluxes containing $MgCl_2$ have been developed (Chesonis and DeYoung, 2008; Courtenay, 2008; Leboeuf et al., 2007):

$$MgCl_2 + 2\underline{Na} = \underline{Mg} + 2NaCl \text{ (slag)} \qquad (12.19)$$

$$MgCl_2 + \underline{Ca} = \underline{Mg} + CaCl_2 \text{ (slag)} \qquad (12.20)$$

These fluxes are premelted with KCl (to prevent hydration of the $MgCl_2$) at one of several compositions corresponding to eutectics in the KCl–$MgCl_2$ system. The price of KCl is volatile, so Courtenay (2011) has demonstrated that satisfactory sodium removal can be achieved with an $MgCl_2$–KCl–$NaCl$ eutectic flux that reduces the cost. Adding 1–3 wt.-% fluorspar (CaF_2) improves reaction efficiency.

Many of the same methods for improving degassing kinetics also improve the kinetics of reactive-metal removal. The effectiveness of chlorine injection is improved by decreasing the bubble size (increasing relative surface area), stirring the melt (to eliminate transport to the bubble surface as a limiting factor), and to a limited extent increasing the concentration of chlorine in the gas. The effective use of flux for calcium and sodium removal requires small particle size and some agitation as well (Chesonis and DeYoung, 2008; Courtenay, 2008; Leboeuf et al., 2007). The concentration of $MgCl_2$ in the flux does not seem to affect removal efficiency.

Inclusions: There are three ways by which inclusions can be removed from aluminum. The first is *sedimentation,* the natural settling of inclusions to the bottom of a melting or holding furnace (Altenpohl, 1998; Engh, 1992; Mirgaux et al., 2008). The rate at which small particles settle in a fluid is governed by a version of Stokes' law:

$$u_r = \frac{2\Delta\rho g a^2}{\rho \nu 9} \qquad (12.21)$$

where

u_r is the settling rate of the inclusion, in m/s

$\Delta\rho$ is the difference between the density of the inclusion and the fluid (in this case, molten aluminum)

g is the gravitational constant (9.81 m/s^2)

a is the inclusion diameter, in m

ρ and ν are the density and kinematic viscosity of molten aluminum, respectively

For a 100 μm (10^{-4} m) alumina inclusion in molten aluminum, Engh (1992) has calculated a typical settling rate of $1.36 \cdot 10^{-3}$ m/s (8.16 cm/min). The settling rate for inclusions smaller than 100 μm is even slower; a 10 μm inclusion would settle at a rate of 0.082 cm/min. Casting furnace operators allow some time for settling to occur, which reduces the number of large inclusions in the melt (Enright, 2007; Instone et al., 2008; Prillhofer et al., 2009). However, for most inclusions, settling is not a predominant means of removal (Mirgaux et al., 2008).

A more effective means of inclusion removal is *flotation* (Altenpohl, 1998; Enright, 2007), which is caused by the attachment of inclusions to bubbles rising to the melt (such as those used for degassing and reactive-metal removal). Complex models have been developed for estimating the rate at which inclusions are removed by flotation in both continuous and batch reactors (Engh, 1992). The removal rate is a function of the following:

- Inclusion size: smaller inclusions are more likely to be floated out (Mirgaux et al., 2009).
- The gas flow rate per unit volume of furnace: more gas equals more bubbles for inclusions to attach to.
- The furnace height: the greater the distance bubbles have to rise, the more likely they are to run into inclusions.
- A *collision efficiency,* equal to the square of the ratio of the inclusion diameter to the bubble diameter.
- The bubble size: smaller bubbles provide more relative surface area.
- The gas inside the bubble: inclusions attach more readily to bubbles containing some chlorine (Enright, 2007; Williams et al., 2000).

The use of Computational fluid dynamics (CFD) to study the impact of process variables on flotation and sedimentation rates during degassing is a major advance in process development (Mirgaux et al., 2008, 2009).

It has been suggested that flotation alone is sufficient to remove enough inclusions to meet quality specifications for some applications. However, in most cases, additional purification is required, especially for smaller inclusions.

The most significant tool for removal of smaller inclusions is *filtration,* which has been used for several decades. There are two main types of industrial filtration (Altenpohl, 1998; Engh, 1992; Enright, 2007): *cake* and *depth.* A cake filter relies on pore openings smaller than the solid inclusions to stop inclusions at the surface, eventually piling them up into a cake of increasing thickness. Cake filters are highly effective, but the metallostatic pressure required to force metal through the cake makes their use impracticable and too expensive. As a result, cake filters are rarely used for filtering molten aluminum, except in very small-volume foundry applications (Pyrotek, 2006).

Figure 12.3 illustrates depth filtration (Aubrey and Smith, 1999; Engh, 1992). Depth filtration relies on the attachment of solid particles to the inside walls of the filter, rather than blocking them at the surface. This prevents a cake from forming at the filter surface, minimizing the pressure-drop problem of cake filters. However, this happens only if the solid particles are much smaller than the pores of the filter. As a result, depth filtration depends upon prior removal of the largest inclusions, usually by flotation. A dual-filter arrangement is often used, in which a large-pore filter is backed by a smaller-pore filter.

Depth filtration is designed primarily to remove inclusions of 5–20 µm diameter. The efficiency of these filters depends primarily on the following (Engh, 1992):

- Inclusion size (larger inclusions are easier to capture)
- Relative filter surface area per unit volume of melt
- Tortuosity (the fraction of filter surface facing the direction of metal flow)
- Filter height
- Metal flow velocity (lower is preferred, in part because higher flow rates cause reentrainment of deposited particles)

Metal flow direction

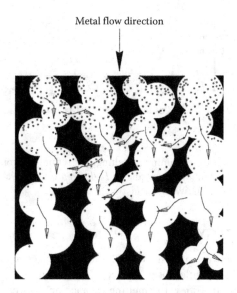

FIGURE 12.3 Mechanism for depth filtration. (From Aubrey, L.S. and Smith, D.D., Technical update on dual stage ceramic foam filtration technology, in *Sixth Australasian-Asian Pacific Conference on Aluminum Cast House Technology*, Whiteley, P.R. and Grandfield, J.F., Eds., TMS-AIME, Warrendale, PA, 1999. With permission.)

The difficulty with maximizing filtration efficiency is designing a more effective filter without also increasing the required metallostatic head beyond an acceptable limit. Several solutions have been devised and will be described later in this chapter.

A significant research effort has been made over the past decade to apply electromagnetic fields for the removal of inclusions from molten aluminum. These fields exert a force on inclusions in the melt, defined by (Taniguchi et al., 2005)

$$F_p = -\frac{3}{2}\frac{\sigma_L - \sigma_p}{2\sigma_L + \sigma_p}V_p F \tag{12.22}$$

where
   F is the applied electromagnetic force
   σ_L and σ_p are the electrical conductivities of molten aluminum and the inclusion, respectively
   V_p is the volume of a hypothetical spherical inclusion

The exerted force is greatest for nonconductive inclusions such as alumina. Several investigators have demonstrated the ability of electromagnetic filtration to remove even small inclusions from molten alumina (Damoah and Zhang, 2012; Yoon et al., 2007); making this technology compatible with existing melting and holding furnaces is a more significant challenge. A recent approach that may be more practical is the application of electromagnetic fields to a depth filter to improve filtration effectiveness (Fritzsch et al., 2013).

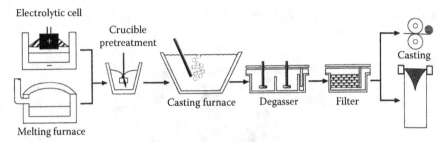

FIGURE 12.4 Flowsheet for aluminum melting and refining. (From Waite, P., Technical perspective on molten aluminum processing, in *Light Metals 2002*, Schneider, W., Ed., TMS-AIME, Warrendale, PA, 2002. With permission.)

REFINING STRATEGY

Figure 12.4, from the work of Waite (2002), illustrates a general flowsheet for molten aluminum processing. While not every recycling facility uses all five of the steps shown here (in some facilities, the melting furnace and the casting furnace are the same), the refining functions described earlier are carried out in one or more of the vessels shown here. The figure points out that the refining sequence for primary metal from electrolytic cells is similar to that for secondary metal from remelting furnaces. As previously mentioned, the refining principles and technology are similar as well.

The development of refining technology over the past 25 years has focused on achieving several goals:

- Reducing alkali metal content, to 3 ppm or less
- Reducing inclusion levels (particularly 5–20 μm) to part-per-trillion levels
- Reducing hydrogen levels to 0.20 cm³/100 g or less
- Decreasing processing times
- Encouraging continuous instead of batch processing
- Decreasing or eliminating the use of chlorine gas

Melting furnace: The primary refining step performed in melting furnaces is demagging (Velasco and Nino, 2011). This requires the use of chlorine gas, as the demagging fluxes described previously are different in composition from the melting fluxes used for absorbing oxide impurities. Previous chlorine-based demagging techniques were inefficient and generated excessive levels of fugitive chlorine gas and chloride fumes. Gas injection pumps are more efficient and are now used extensively for demagging (Pyrotek, 2006). They may also be useful for inclusion flotation and alkali removal, although not to the levels required for product quality purposes. The fluoride-based fluxes used for dross removal during melting also help with magnesium and other alkali removal, as per reactions (12.16 through 12.18).

Crucible pretreatment: The use of crucible pretreatment is not universal, since improvements in other refining technologies appear to have made it less necessary. However, it may still have advantages, especially for alkali removal. Haugen et al.

(2011) have demonstrated that alkali removal in the crucible can reduce dross formation in the casting furnace by 25%–40%. One advantage of crucible treatment is that it allows the use of solid fluxes as a replacement for chlorine gas. These fluxes contain $MgCl_2$ to carry out reactions (12.19) and (12.20); the $MgCl_2$ is contained in a premelted mixture with KCl and NaCl, as described previously (Riverin et al., 2006). Some crucible treatment processes use AlF_3 instead. The development of a rotary flux injector (Figure 12.5) has improved the efficiency of solid flux injection over the previous lancing technique.

Casting furnace: In many facilities, the melting and casting furnaces are the same, and so the discussion of refining practice in melting furnaces from Chapter 9 applies here. If there is no crucible treatment facility, the casting furnace is where alkali removal and primary degassing take place. As Haugen et al. (2011) point out, this presents a problem. Reverberatory melting and casting furnaces are designed with heat transfer in mind and do not function well as chemical reactors. In addition, the time required to complete refining in the furnace lowers productivity.

Refining in casting furnaces was traditionally performed using stationary lances injecting a mixture of nitrogen and chlorine gas. The large bubble size and poor mixing of these lances led to low effectiveness, wasted chlorine, and workplace hazards caused by the unreacted chlorine (Waite, 2002). As a result, the development of better mixing technology is important in casting-furnace refining. Rotary gas injectors (RGIs) were introduced in the 1970s and have largely replaced the lances (Chesonis et al., 2009). These devices feature a spinning rotor with an impeller at the end that breaks up gas bubbles as they are injected into the melt. Since their introduction, the development of improved impeller designs has become a small industry in itself (Mi et al., 2008); CFD simulation plays a major role in the process. Porous plugs have also been used as a gas delivery device (Gamweger and Schweiger, 2007); the finer bubble size that these devices deliver shortens degassing time but eliminates the possibility of simultaneous fluxing.

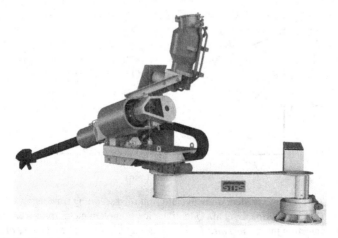

FIGURE 12.5 A pivoting rotary flux injector for molten aluminum refining. (Photo courtesy of Marc-André Thibault, STAS.)

Figures 12.6 through 12.8, from the results of Hopkins et al. (1995), show the results achievable in a casting furnace using a rotary gas injection pump. Figure 12.6 shows the hydrogen content of molten aluminum (~2.5% Mg) under different refining conditions. As previously mentioned, the use of chlorine in the purge gas makes no difference in the reduction of hydrogen content. Figure 12.7 shows the inclusion content of the melt as a function of time; here, the impact of adding 5% Cl_2 to the gas is obvious. Figure 12.8 shows the reduction of alkali content achievable when

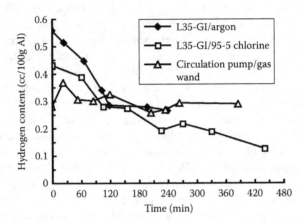

FIGURE 12.6 Hydrogen content of molten aluminum vs. refining time using Ar and Ar–5% Cl_2 mixture. (From Hopkins, L. et al., Quantification of molten metal improvements using an L-series gas injection pump, in *Third International Symposium on Recycling of Metals and Engineered Materials*, Queneau, P.B. and Peterson, R.D., Eds., TMS-AIME, Warrendale, PA, 1995. With permission.)

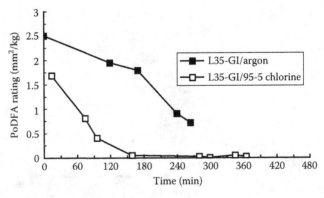

FIGURE 12.7 Inclusion content in molten aluminum vs. refining time using Ar and Ar–5% Cl_2 mixture. (From Hopkins, L. et al., Quantification of molten metal improvements using an L-series gas injection pump, in *Third International Symposium on Recycling of Metals and Engineered Materials*, Queneau, P.B. and Peterson, R.D., Eds., TMS-AIME, Warrendale, PA, 1995. With permission.)

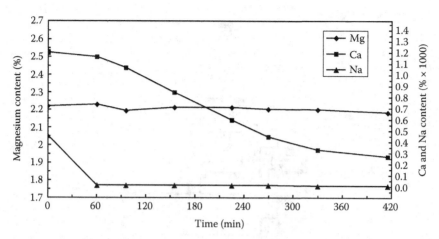

FIGURE 12.8 Magnesium and molten-alkali content of molten aluminum vs. refining time using Ar–5% Cl_2 mixture. (From Hopkins, L. et al., Quantification of molten metal improvements using an L-series gas injection pump, in *Third International Symposium on Recycling of Metals and Engineered Materials*, Queneau, P.B. and Peterson, R.D., Eds., TMS-AIME, Warrendale, PA, 1995. With permission.)

using an Ar–5% Cl_2 mixture. As a result of reactions (12.19) and (12.20), most of the calcium and sodium are removed without removing the magnesium as well. As the figures show, casting-furnace refining is effective, especially at alkali and inclusion removal, but very slow and inefficient. In addition, metal that has been degassed in the casting furnace regains its hydrogen content quickly while waiting to be cast (Enright, 2007). Because of this, the trend over the past generation has been to minimize degassing in the casting furnace and replace it with *in-line* degassing and filtration (Fielding, 1996; Leboeuf et al., 2007). In-line processes are designed specifically for hydrogen and inclusion removal. Their increased efficiency and kinetics justify the capital and operating costs of the additional processing (Williams et al., 2000).

In-line degassing: Figure 12.9 shows a cross section of the Spinning nozzle inert flotation (SNIF) process, one of several in-line degassing devices introduced in the early 1980s. The device features a spinning (300–400 rpm) rotor inserted in the molten aluminum, through which a gas mixture (often Ar–5% Cl_2) is fed. The spinning rotor breaks up the gas into very fine bubbles, which increase the relative amount of surface area between the metal and the gas. This improves degassing kinetics and efficiency. The smaller bubbles also encourage removal of smaller inclusions. In-line devices like this are designed to operate continuously, with molten aluminum constantly being fed at the top and removed through a plug in the bottom. The vessel size is determined by the metal flow-through rate and the time required to lower the hydrogen level to a desired value.

Since their introduction, a number of improvements have been made to in-line degassing units like this (Waite, 2002). The use of multiple stages generates lower hydrogen levels than a single unit of comparable size. The degassers are now sealed to prevent contact with air (Le Roy and Menet, 2007) and flushed with argon to

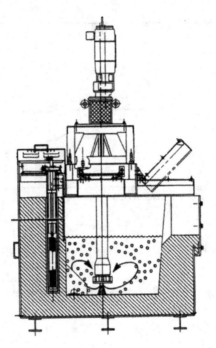

FIGURE 12.9 Schematic of the SNIF degassing vessel. (From Altenpohl, D.G., *Aluminum: Technology, Applications, and Environment*, 6th edn., TMS-AIME, Warrendale, PA, 1998. With permission.)

reduce dross generation. Faster rotor speed (900 rpm in new Hycast units) results in even smaller bubble size and better kinetics (Steen et al., 2010). The most recent Alpur and Hycast devices include equipment for heating the metal, which is important if molten metal is to be held in the unit between casts. Improved drainage has also been a priority, making it easier to change alloys with less wasted metal.

The newest trend in in-line degassing eliminates the degassing vessel altogether. Figure 12.10 illustrates the operating principle behind the Alcan Compact Degasser (ACD), which substitutes the casting trough for the degassing vessel (Lavoie et al., 1996; Maltais et al., 2008). Baffles located in the trough create the equivalent of a

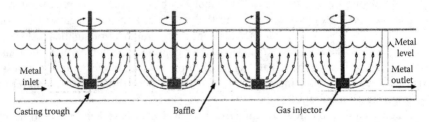

FIGURE 12.10 Schematic of the Alcan Compact Degasser. (From Waite, P. et al., in *Fifth Australasian-Asian Pacific Conference on Aluminum Cast House Technology*, Nilmani, M., Whiteley, P., and Grandfield, J., Eds., TMS-AIME, Warrendale, PA, 1997. With permission.)

multistage degassing vessel, with a rotary gas injector (RGI) for each compartment. Degassing and inclusion removal efficiencies in this device are similar to that of a degassing vessel; however, the use of the trough reduces the space requirement considerably. In addition, use of a trough degasser eliminates metal retention between casts (Fielding, 1996). This is important in plants casting several alloys, since metal retained in a degassing vessel between casts must usually be drained and remelted. The most recent version of the ACD is also sealed, which reduces dross formation.

As previously mentioned, chlorine gas has traditionally been used to remove alkaline impurities (Ca, Li, Na) from the melt. It also improves flotation of inclusions. However, workplace hygiene concerns and the regulatory environment have provided increasing incentive for casthouses to adopt *chlorine-free* processing methods (Chesonis et al., 2009). This is done using equipment that injects $MgCl_2$-containing fluxes in place of the Cl_2, using Ar as the carrier. The rotary flux injector (RFI) replaces the RGI (Ohno, 2010); the Hycast™ Ram replaces the Hycast I-60 (Haugen et al., 2011); the Salt-ACD replaces the ACD (Robichaud et al., 2011). Performance in alkali removal and degassing is at least as good as that with chlorine gas injection; inclusion removal may be even better (Maltais et al., 2008). It is expected that the use of chlorine gas will disappear over time.

Filtration: Filtration of molten secondary aluminum has been practiced for several decades but has in recent years become increasingly important as product quality standards have tightened (Waite, 2002). As previously mentioned, almost all filtration technology in molten aluminum uses the depth principle. However, numerous filter materials and filtration devices exist within that category (Weaver, 1997). The choice of filtration technology depends on a number of factors, including the following:

- *Type of product*: Castings have a higher tolerance for inclusions than wrought product and may not require the same degree of filtration.
- *Plant production rate*: Melting facilities with higher capacity can justify investing in more expensive types of filtration equipment.
- *Number of alloys produced*: A facility that frequently changes alloy will avoid filters that retain a large volume of metal between casts.
- *Available floor space*: Some filtration equipment has a much larger footprint than others.
- *Type of inclusion*: Most of the inclusions present in remelted scrap consist of oxides, which interact better with some filter materials than others. However, the use of $MgCl_2$-based fluxes for alkali removal creates concern over the presence of small $MgCl_2$ inclusions in the metal (DeYoung, 1999). Removing these requires different filter materials.

Weaver (1997) divides the filters used for molten aluminum into two groups: those meant for a single use and those that can be used for several times before replacement. Multiple-use filters have a higher capital cost and a larger footprint. They also have higher capacities and lower operating costs, since they are replaced less often. As a result, they are more common in large-volume shops casting a limited number of alloys. Figure 12.11 illustrates the granular type of *deep-bed filter* (DBF),

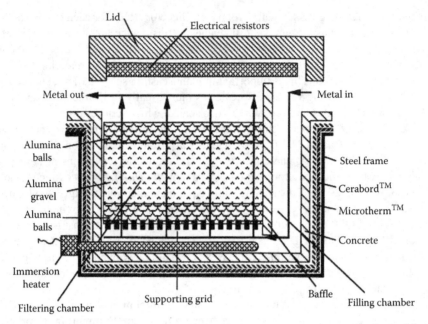

FIGURE 12.11 Construction of a granular deep-bed filter. (From Clement, G., The Pechiney deep bed filter: Technology and performance, in *Light Metals 1995*, Evans, J., Ed., TMS-AIME, Warrendale, PA, 1995. With permission.)

which has been in use since the 1940s (Clement, 1995; DeYoung, 1999; Fielding, 1996). Metal is fed through the bottom of the filter to reduce turbulence and flows upward through a composite structure consisting of alumina balls (13–19 mm dia.) and gravel (3–6 mesh). A typical flow rate through the filter is 0.1–0.4 cm/s; the low velocity improves filtration efficiency. The supporting grid at the bottom distributes the metal evenly and reduces channeling. Heaters are used to maintain metal temperature in the filter. Top-flow DBFs are also available (Le Roy et al., 2007). DBFs are more effective than other types of filter but require flushing to avoid off-spec metal during alloy changes.

Figure 12.12 shows a more recent type of multiple-use filter (Pyrotek, 2008), the *rigid-media tube filter* (*RMTF*), also known as a *bonded-particle filter*. The cartridge in the figure is a bundle of tubes made up of bonded ceramic particles (usually silicon carbide) with variable-size pores. Metal fed into the filter flows through the pores into the inside of the tubes and from there to the outlet. The RMTF is the most efficient filter available, but breakage of the tubes is a concern (DeYoung, 1999), and as a result its use in this form is not widespread.

Introduced in the 1970s, the single-use foam filter has become the most widely used in the industry (Pyrotek, 2008). Figure 12.13 shows several shapes of ceramic foam filter (CFF), which is produced by impregnating porous polyurethane foam with a ceramic slurry, allowing the slurry to dry, and firing the foam (Chesonis et al., 2002; Parker et al., 1999). When the organic foam decomposes and vaporizes, the sintered ceramic structure left behind forms a framework capable of filtering

FIGURE 12.12 A rigid-media tube filter. (From Eichenmiller, D.J. et al., in *Light Metals 1994*, Mannweiler, U., Ed., TMS-AIME, Warrendale, PA, 1994. With permission.)

FIGURE 12.13 Different shapes of foam filter. (From Parker, G. et al., Production scale evaluation of a new design ceramic foam filter, in *Light Metals 1999*, Eckert, C.E., Ed., TMS-AIME, Warrendale, PA, p. 1057, 1999. With permission.)

molten metal. CFFs can be produced with a range of pore sizes and cost less than other filters (DeYoung, 1999). However, they are less efficient than other filter materials (Fielding, 1996), limiting their usefulness. Figure 12.14 shows a common arrangement, a staged CFF (Aubrey and Smith, 1999; Chesonis et al., 2002). In this arrangement, a *coarse* filter occupies the top slot in the filter frame, and a finer-pore filter the lower position. This improves capacity and improves capture efficiency for smaller inclusions.

Considerable efforts have been made in recent years to improve CFFs. The choice of ceramic material has been a topic of particular concern. CFFs are typically produced from slurries of phosphate-bonded alumina or silicon carbide (Foseco, 1998).

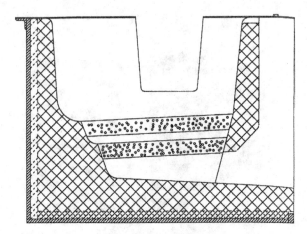

FIGURE 12.14 Cross section of a staged ceramic foam filter. (From Aubrey, L.S. and Smith, D.D., Technical update on dual stage ceramic foam filtration technology, in *Sixth Australasian-Asian Pacific Conference on Aluminum Cast House Technology*, Whiteley, P.R. and Grandfield, J.F., Eds., TMS-AIME, Warrendale, PA, 1999. With permission.)

The more common alumina filters are brittle and tend to break in service, introducing large inclusions to the melt. In addition, the phosphate used for bonding can react over time with the molten aluminum, resulting in undesirable levels of phosphorus dissolved in the metal. Spent filters can react with atmospheric moisture, generating hazardous phosphine gas (Aubrey et al., 2010). Silicon carbide filters can also break in service and occasionally *contribute* silicon to low-Si alloys. Recent innovations include the following:

- Using a mixture of metallic aluminum powder and alumina in the slurry, which produces a finer-grained structure when fired (Luyten et al., 2006)
- Incorporating SiC and MgO in the slurry, which improves compressive strength and thermal shock resistance (Cao et al., 2007)
- Replacing the alumina with graphite (Foseco, 1998) or boron glass–bonded mullite (Aubrey et al., 2010)
- Coating the filter interior with molten flux (Ni et al., 2006), which seems to improve capture of inclusions

The bonded-particle filter material described earlier can be manufactured in disposable elements as well as tube cartridges. The elements are reusable, resulting in cost savings over single-use ceramic foam, and their higher efficiencies make them useful for applications requiring better removal of finer inclusions (Keegan et al., 1996). Other innovations include the development of multistage filters designed to remove liquid $MgCl_2$ as well as solid inclusions and the production of filter materials for improved removal of nonoxide inclusions.

Figure 12.15, adapted from Enright (2007), demonstrates how degassing and CFF can reduce the concentration of inclusions in molten aluminum. The inclusions left after filtration consist mostly of small carbides and MgO, and the oxide film content

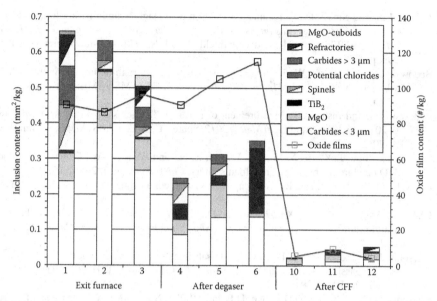

FIGURE 12.15 Impact of degassing and ceramic foam filtration on concentration of inclusions in molten secondary aluminum. (After Enright, P.G., Molten metal treatment technologies for secondary aluminium processing, presented at *9th OEA Recycling Congress*, Cologne, Germany, 2007.)

has been reduced by 90%. Finding a way of removing the small carbides will be a continuing challenge in molten aluminum purification.

RECOMMENDED READING

Altenpohl, D.G., *Aluminum: Technology, Applications, and Environment*, 6th edn., TMS-AIME, Warrendale, PA, 1998, p. 61.

Engh, T.A., *Principles of Metal Refining*, Oxford University Press, Oxford, U.K., 1992, Chapters 4 and 5.

Fielding, R.A.P., The role of grain refining, degassing and filtration in the production of quality ingot products, *Light Metal Age*, 54(8), 46, October 1996.

Waite, P., Technical perspective on molten aluminum processing, in *Light Metals 2002*, Schneider, W., Ed., TMS-AIME, Warrendale, PA, 2002, p. 841.

REFERENCES

Altenpohl, D.G., *Aluminum: Technology, Applications, and Environment*, 6th edn., TMS-AIME, Warrendale, PA, 1998, p. 61.

Apelian, D., *Aluminum Cast Alloys: Enabling Tools for Improved Performance*, NADCA, Wheeling, IL, July 21, 2009. http://www.diecasting.org/research/wwr/WWR_AluminumCastAlloys.pdf.

Aubrey, L.S., Olson, R., and Smith, D.D., Development of a phosphate–free reticulated foam filter material for aluminium cast houses, *Mater. Sci. Forum*, 630, 137, 2010.

Aubrey, L.S. and Smith, D.D., Technical update on dual stage ceramic foam filtration technology, in *Sixth Australasian-Asian Pacific Conference on Aluminum Cast House Technology*, Whiteley, P.R. and Grandfield, J.F., Eds., TMS-AIME, Warrendale, PA, 1999, p. 133.

Badowski, M. and Droste, W., Hydrogen measurement practices in liquid aluminium at low hydrogen levels, in *Light Metals 2009*, Bearne, G., Ed., TMS-AIME, Warrendale, PA, 2009, p. 701.

Badowski, M. and Instone, S., Measurement of non-metallic inclusions in the size range of 10–20 μm by LiMCA, in *Light Metals 2012*, Suarez, C.E., Ed., TMS-AIME, Warrendale, PA, 2012, 1077.

Cao, D.-L., Zhou, J.-Y., Jin, Y.-M., Ma, L., Shi, Z.-N., and Wang, Z.-W., Effects of Sic and MgO on alumina–based ceramic foams filters, *China Foundry*, 4, 292, 2007.

Carey, R., NASAAC—An analysis of a capitalist system gone wrong, *Die Casting Eng.*, 56(1), 36, January 2013.

Chen, C., Wang, J., Shu, D., Xue, J., Sun, B.-D., Xue, Y.-S., and Yan, Q.-M., Iron reduction in aluminum by electroslag refining, *Trans. Nonferrous Met. Soc. China*, 22, 964, 2012.

Chesonis, D.C. and DeYoung, D.H., Chloride salt injection to replace chlorine in the Alcoa A622 degassing process, in *Light Metals 2008*, DeYoung, D.H., Ed., TMS-AIME, Warrendale, PA, 2008, p. 569.

Chesonis, D.C., DeYoung, D.H., Lake, D.E., and Ridler, N.R., LiMCA comparison of a bed filter and a two stage ceramic foam filter, in *Light Metals 2002*, Schneider, W., Ed., TMS-AIME, Warrendale, PA, 2002, p. 937.

Chesonis, D.C., Williams, E.M., and DeYoung, D.H., Meeting environmental challenges in the casthouse, in *Light Metals 2009*, Bearne, G., Ed., TMS-AIME, Warrendale, PA, 2009, p. 653.

Clement, G., The Pechiney deep bed filter: Technology and performance, in *Light Metals 1995*, Evans, J., Ed., TMS-AIME, Warrendale, PA, 1995, p. 1253.

Courtenay, J.H., Improved understanding of the melting behaviour of fused magnesium chloride–potassium chloride based refining fluxes, in *Light Metals 2008*, DeYoung, D.H., Ed., TMS-AIME, Warrendale, PA, 2008, p. 637.

Courtenay, J.H., Development of a fused magnesium chloride containing refining flux based on a ternary system, *Mater. Sci. Forum*, 693, 161, 2011.

Damoah, L.N.W. and Zhang, L., High frequency electromagnetic separation of inclusions from aluminum, in *Light Metals 2012*, Suarez, C.E., Ed., TMS-AIME, Warrendale, PA, 2012, p. 1069.

Dewan, M.A., Rhamdhani, M.A., Mitchell, J.B., Davidson, C.J., Brooks, G.A., Easton, M., and Grandfield, J.F., Control and removal of impurities from Al melts: A review, *Mater. Sci. Forum*, 693, 149, 2011.

DeYoung, D.H., Metal filtration performance: Removal of molten salt inclusions, in *Sixth Australasian-Asian Pacific Conference Aluminum Cast House Technology*, Whiteley, P.R. and Grandfield, J.F., Eds., TMS-AIME, Warrendale, PA, 1999, p. 121.

Eckert, C.E., The origin and identification of inclusions in foundry alloy, in *Third International Conference Molten Aluminum Processing*, AFS, Des Plaines, IL, 1992, p. 17.

Eichenmiller, D.J., Henderson, R.S., and Neff, D.V., Rigid media filtration: New understanding and possibilities with bonded particle filters, in *Light Metals 1994*, Mannweiler, U., Ed., TMS-AIME, Warrendale, PA, 1994.

Engh, T.A., *Principles of Metal Refining*, Oxford Science Publications, Oxford, U.K., 1992.

Enright, P.G., Molten metal treatment technologies for secondary aluminium processing, presented at *9th OEA Recycling Congress*, Cologne, Germany, February 26, 2007.

Exebio, J.M., Larouche, D., Paquin, D., Proulx, J., and Dupuis, C., Investigation of hydrogen measurement technique for molten aluminum, *Light Metals 2008*, DeYoung, D.H., Ed., TMS-AIME, Warrendale, PA, 2008, p. 749.

Fielding, R.A.P., The role of grain refining, degassing and filtration in the production of quality ingot products, *Light Metal Age*, 54(8), 46, October 1996.

Foseco, Inc., The application of foam filters to optimize aluminum casting production, November 28, 1998. http://www.foseco.com.tr/tr/downloads/FoundryPractice/227-02_Application_of_sivex_foam_filters.pdf.

Foseco, Inc., The technology of batch degassing for hydrogen removal from aluminium melts utilizing different rotor designs, September 1, 2011. http://www.foseco.se/se/press_room/pdf/FP_256/FP256-03.pdf.

Fritzsch, R., Kennedy, M.W., Akhtar, S., Bakken, J.A., and Aune, R.E., Electromagnetically modified filtration of liquid aluminium with a ceramic foam filter, in *Light Metals 2013*, Sadler, B., Ed., TMS-AIME, Warrendale, PA, 2013. http://www.metallurgy.no/A10%20.pdf.

Fruehan, R. and Anyalebechi, P., Gases in metals, in *ASM Handbook, Vol. 15: Casting*, ASM International, Materials Park, OH, 2008, p. 64.

Gamweger, K. and Schweiger, R., Porous plug technology for degassing in aluminum foundry ladles, in *Light Metals 2007*, Sørlie, M., Ed., TMS-AIME, Warrendale, PA, 2007, p. 635.

Gansemer, T.M., Reynolds, B.C., Hart, J.N., and Chesonis, D.C., Improvement in hydrogen measurement technique for molten aluminum, in *Light Metals 2007*, Sørlie, M., Ed., TMS-AIME, Warrendale, PA, 2007, p. 685.

Gao, J.W., Shu, D., Wang, J., and Sun, B.D., Study on iron purification from aluminium melt by $Na_2B_4O_7$ flux, *Mater. Sci. Technol.*, 25, 619, 2009.

Hamouche, H., *Trace Element Analysis in Aluminium Alloys*, Alcan International Ltd., May 14, 2004. http://www.armi.com/News/Presentations/2003/Alcan%20Presentation.pdf.

Haugen, T., Steen, I.K., Myrbostad, E., and Håkonsen, A., Crucible fluxing with Hycast RAM—Effect on metal quality and operational cost, *Mater. Sci. Forum*, 693, 44, 2011.

Hills, M.P., Thompson, C., Henson, M.A., Moores, A., Schwandt, C., and Kumar, R.V., Accurate measurement of hydrogen in molten aluminium using current reversal mode, in *Light Metals 2009*, Bearne, G., Ed., TMS-AIME, Warrendale, PA, 2009, p. 707.

Hopkins, L., Beasley, J., Henderson, R.S., and Campbell, P.S., Quantification of molten metal improvements using an L-series gas injection pump, in *Third International Symposium on Recycling of Metals and Engineered Materials*, Queneau, P.B. and Peterson, R.D., Eds., TMS-AIME, Warrendale, PA, 1995, p. 31.

Instone, S., Buchholz, A., and Gruen, G.-U., Inclusion transport in casting furnaces, in *Light Metals 2008*, DeYoung, D.H., Ed., TMS-AIME, Warrendale, PA, 2008, p. 811.

Keegan, N.J., Schneider, W., Krug, H.P., Moritz, J., Stolz, M., and Dopp, V., Evaluating ceramic–foam and bonded–particle cartridge filtration systems, *JOM*, 48(8), 28, August 1996.

Kurban, M., Sommerville, I.D., Mountford, N.D.G., and Mountford, P.H., An ultrasonic sensor for the continuous monitoring of the cleanliness of liquid aluminum, in *Light Metals 2005*, Ed., TMS-AIME, Warrendale, PA, 2005, p. 945.

Lavoie, S., Pilote, E., Thibault, M.A., and Pomerleau, C.J., The Alcan Compact Degasser, a trough-based aluminum treatment process. Part II: Equipment description and plant experience, in *Light Metals 1996*, Hale, W., Ed., TMS-AIME, Warrendale, PA, 1996, p. 1007.

Le Roy, G., Chateau, J.-M., and Charlier, P., PBDF; Proven Filtration for high-end applications, in *Light Metals 2007*, Sørlie, M., Ed., TMS-AIME, Warrendale, PA, 2007, p. 651.

Le Roy, G. and Menet, P.-Y., Improved rotor design in a totally sealed degasser, *Alum. Int. Today*, 19(3), 26, May/June 2007.

Leboeuf, S., Dupuis, C., Maltais, B., Thibault, M.-A., and Smarason, E., In-line salt fluxing process: The solution to chlorine gas utilization in casthouses, in *Light Metals 2007*, Sørlie, M., Ed., TMS-AIME, Warrendale, PA, 2007, p. 623.

Lee, G.-C., Kim, M.-G., Park, J.-P., Kim, J.-H., Jung, J.-H., and Baek, E.-R., Iron removal in aluminum melts containing scrap by electromagnetic stirring, *Mater. Sci. Forum*, 638–642, 267, 2010.

Logan Aluminum, Inc., *Chemistry and Chemical Sampling: Updated 2012*, February 27, 2012. http://www.loganrawmaterials.com/pdf/Chemistry%20and%20Chemical%20Sampling%20-%20Web%20version.pdf.

Luyten, J., Vanderneulen, W., De Schutter, F., Simensen, C., and Ryckeboer, M., Ceramic foams for Al-recycling, *Adv. Eng. Mater.*, 8, 705, 2006.

Maltais, B., Privé, D., Taylor, M., and Thibault, M.-A., Metal treatment update, in *Light Metals 2008*, DeYoung, D.H., Ed., TMS-AIME, Warrendale, PA, 2008, p. 547.

Mi, G., Liu, X., Wang, K., Qi, S., Wang, H., and Niu, J., Analyses of the influencing factors of rotating impeller degassing process and water simulation experiment, *Mater. Sci. Forum*, 575–578, 1258, 2008.

Mirgaux, O., Ablitzer, D., Waz, E., and Bellot, J.P., Removal of inclusions from molten aluminium by flotation in a stirred reactor: A mathematical model and a computer simulation, *Int. J. Chem. Reactor Eng.*, 6, A52, 2008.

Mirgaux, O., Ablitzer, D., Waz, E., and Bellot, J.P., Mathematical modeling and computer simulation of molten aluminum purification by flotation in stirred reactor, *Metall. Mater. Trans B.*, 40B, 363, 2009.

Moores, A., A new device for determining H_2 in aluminium alloys, *Aluminium*, 84(7), 42, July/August 2008.

de Moraes, H.L., de Oliveira, J.R., Espinosa, D.C.R., and Tenorio, J.A.S., Removal of iron from molten recycled aluminum through intermediate phase filtration, *Mater. Trans.*, 47, 1731, 2006.

Neff, D.V., Evaluating molten metal cleanliness for producing high integrity aluminum die castings, *Die Casting Eng.*, 48(9), 24, September 2004.

Ni, H., Yu, Z., Mingyu, H., Weibiao, G., and Sun, B., Purifying effects and mechanism of a new composite filter, *Mater. Sci. Eng. A*, 426, 53, 2006.

Ohno, Y., The latest melt refining technology in furnace for environmental improvement, in *12th International Conference on Aluminum Alloys*, Kumai, S., Ed., The Japan Institute of Light Metals, Tokyo, Japan, 2010. http://www.pyrotek.info/documents/techpapers/Melt_Refining-Environmental_Improvement.pdf.

Parker, G., Williams, T., and Black, J., Production scale evaluation of a new design ceramic foam filter, in *Light Metals 1999*, Eckert, C.E., Ed., TMS-AIME, Warrendale, PA, 1999, p. 1057.

Pascual, A., Jr., Emerging melt quality control solution technologies for aluminium melt, *China Foundry*, 6(4), 358, November 2009.

Peterson, R.D., Wells, P.A., and Creel II, J.M., Reducing chlorine usage in furnace fluxing: Two case studies, in *Light Metals 1995*, Evans, J., Ed., TMS-AIME, Warrendale, PA, 1995, p. 1197.

Poynton, S., Brandt, M., and Grandfield, J., A review of inclusion detection methods in molten aluminum, in *Light Metals 2009*, Bearne, G., Ed., TMS-AIME, Warrendale, PA, 2009, p. 681.

Poynton, S., Brandt, M., and Grandfield, J., The use of electromagnetic fields for the detection of inclusions in aluminium, *Mater. Sci. Forum*, 630, 155, 2010.

Prillhofer, B., Antrekowitsch, H., Böttcher, H., and Enright, P., Nonmetallic inclusions in the secondary aluminum industry for the production of aerospace alloys, in *Light Metals 2008*, DeYoung, D.H., Ed., TMS-AIME, Warrendale, PA, 2008, p. 603.

Prillhofer, B., Böttcher, H., and Antrekowitsch, H., A new methodology for performance evaluation of melt refinement process in the aluminum industry, in *Light Metals 2009*, Bearne, G., Ed., TMS-AIME, Warrendale, PA, 2009, p. 689.

Pyrotek, Inc., Improving performance in the furnace melt treatment process, January 19, 2006. http://www.pyrotek.info/melt_treatment.

Pyrotek, Inc., Improving performance in the filtration process, August 17, 2008. http://www. pyrotek-inc.com/documents/brochure/843_AIT_Filtration_Insert_web.pdf.

Riverin, G., Bilodeau, J.-F., and Dupuis, C., A novel crucible metal treatment process for impurity removal in secondary aluminium, in *Light Metals 2006*, Galloway, T.J., Ed., TMS-AIME, Warrendale, PA, 2006, p. 759.

Robichaud, P., Dupuis, C., Mathis, A., Côté, P., and Maltais, B., In-line Salt-ACD™: A chlorine–free technology for metal treatment, in *Light Metals 2011*, Lindsay, S.J., Ed., TMS-AIME, Warrendale, PA, 2011, p. 739.

de la Sabloniere, H. and Samuel, F.H., PoDFA measurements of inclusions in 319.1 alloy: Effect of Mg (0.45 wt%) addition and role of sludge, *AFS Trans.*, 96, 751, 1996.

Sigworth, G.K., Williams, E.M., and Chesonis, D.C., Gas fluxing of molten aluminum: An overview, in *Light Metals 2008*, DeYoung, D.H., Ed., TMS-AIME, Warrendale, PA, 2008, p. 581.

Steen, I.K., Myrbostad, E., Hakonsen, A., and Haugen, T., Hycast I-60 SIR—A unique concept for inline melt refining, in *Light Metals 2010*, Schneider, W., Ed., TMS-AIME, Warrendale, PA, 2010, p. 61.

Taniguchi, S., Yoshikawa, N., and Takahashi, K., Application of EPM to the separation of inclusion particles from liquid metal, in *The 15th Riga and 6th PAMIR Conference on Fundamental and Applied MHD*, Alemany, A., Ed., Salaspils Institute of Physics, Riga, Latvia, 2005, p. 55. http://ipul.lv/pamir/cd/vol.I/riga-pamir-vol.I-55.pdf.

Utigard, T., Thermodynamic considerations of aluminum refining and fluxing, in *Extraction, Refining, and Fabrication of Light Metals*, Sahoo, M. and Pinfold, P., Eds., CIM, Montreal, Canada, 1991, p. 353.

Velasco, E. and Montalvo, F., Influence of reduced pressure on density of aluminium extrusion alloy, *Int. J. Cast Met. Res.*, 25, 59, 2012.

Velasco, E. and Nino, J., Recycling of aluminium scrap for secondary Al-Si alloys, *Waste Manage. Res.*, 29(7), 686, 2011.

Waite, P., Technical perspective on molten aluminum processing, in *Light Metals 2002*, Schneider, W., Ed., TMS-AIME, Warrendale, PA, 2002, p. 841.

Waite, P., Lavoie, S., and Pilote, E., The Alcan Compact Degasser: Operational experience and performance, in *Fifth Australasian-Asian Pacific Conference on Aluminum Cast House Technology*, Nilmani, M., Whiteley, P., and Grandfield, J., Eds., TMS-AIME, Warrendale, PA, 1997.

Weaver, C., Future trends in melt quality control, in *Fifth Australasian-Asian Pacific Conference on Aluminum Cast House Technology*, Nilmani, M., Whiteley, P., and Grandfield, J., Eds., TMS-AIME, Warrendale, PA, 1997, p. 105.

Williams, E.M., McCarthy, R.W., Levy, S.A., and Sigworth, G.K., Removal of alkali metals from aluminum, in *Light Metals 2000*, Peterson, R.D., Ed., TMS-AIME, Warrendale, PA, 2000, p. 785.

Yoon, E.P., Choi, J.P., Seo, Y.S., and Nam, T.W., The continuous elimination of inclusions in molten aluminum by direct and alternate electromotive force, *Mater. Sci. Forum*, 539–543, 499, 2007.

Zhang, L. and Damoah, L.N., Current technologies for the removal of iron from aluminum alloys, in *Light Metals 2011*, Lindsay, S.J., Ed., TMS-AIME, Warrendale, PA, 2011, p. 757.

Zhang, L., Lv, X., Torgerson, A.T., and Long, M., Removal of impurity elements from molten aluminum: A review, *Metal. Proc. Extr., Met. Rev.*, 32, 150, 2011.

Zholnin, A.G., Novichkov, S.B., and Stroganov, A.G., Choice of additions to NaCl-KCl mixture for aluminum refining from alkali and alkaline–earth impurities, in *Light Metals 2005*, Kvande, H., Ed., TMS-AIME, Warrendale, PA, 2005, p. 973.

13 Dross Processing

The furnaces used for melting recycled aluminum generate three products. The first is the molten aluminum, which is cast or transferred to customers. The second is the off-gas, which will be discussed in the next chapter. The third is a semi-solid skim removed from the melt surface. This skim, better known as dross, is a mixture of molten aluminum metal and various oxide and chloride compounds. Its composition depends on the choice of melting practice and the amount and composition of fluxes used. Regardless of composition, the processing of dross has always been important. Dross processing can (a) recover the valuable metal and salt content of the dross and (b) minimize the amount of waste material to be disposed of after treatment. This chapter will describe the choices available in dross treatment technology and the factors determining whether one technology is favored over another.

TYPES OF DROSS

In general, drosses fall into two categories: *nonsalt dross* and *salt dross* (Peterson, 2011).

Nonsalt dross (better known as *white* or gray dross), the more common of the two types, is produced by melting facilities that melt without using flux. These include all electric furnaces, reverberatory furnaces melting bulky scrap or ingot, and the holding furnaces used for primary aluminum. It can be generated in four ways (van Linden, 1997):

1. During molten metal transfer, when the oxide skin on the metal surface ruptures, exposing new surface and generating more oxide (Prakash et al., 2011).
2. During melting and holding, when the oxide skin forms on the melt surface. Furnaces that directly expose metal to burner flames tend to produce a lot of dross in this fashion. High-melt-loss types of scrap often have surface oxide present before melting starts and as a result generate more dross at this stage of the recycling process; thin-walled scrap is also more likely to generate dross during melting.
3. During molten metal processing and refining, when stirring and surface disruption create fresh surface for oxidation.
4. As a result of metal spills or *skulls*. These are not drosses themselves but are often added to the dross recovered from the melt surface for remelting.

The amount of dross generated varies according to the type of scrap being melted; the International Aluminium Institute estimates that it ranges from 1% to 5% of the charge for secondary smelters and 1% for primary smelters (Smith, 2008).

Worldwide *production* of dross is estimated at 760,000 ton/year, about half from primary smelters and half from secondary.

The first key to dross *treatment* is reducing the amount generated in the first place. This can best be accomplished by minimizing the amount of surface disruption that takes place during melting. The use of level transfer rather than cascading during metal transfer is helpful, because it limits secondary flows in the holding furnace or ladle after metal is poured from a crucible (Prakash et al., 2011). A more effective tactic is the submergence of scrap during melting to prevent surface disruption; the vortex technology described in Chapter 8 is an example. Keeping flame temperatures down during melting in stationary furnaces (i.e., reverbs) reduces oxidation, as does reducing temperature stratification. Better charging practice to minimize direct metal contact with burner flames also helps.

Figure 13.1 shows the light grey color of nonsalt dross (Urbach, 2011). It consists almost entirely of Al_2O_3 and aluminum metal trapped by the surface tension of the oxide skin. (Because of the amount of molten aluminum they contain, these drosses are sometimes called *wet*.) The metal content can vary from 15% to 80%, depending on the amount of melt loss and the sampling technique. Small amounts of aluminum carbide (Al_4C_3) and aluminum nitride (AlN) are also frequently present (Peterson, 2011), caused by reactions that occur mostly after the dross has been removed from the furnace. Dross from primary smelting furnaces may also contain small levels of cryolite (Na_3AlF_6). This is the result of electrolytic cell bath accidentally being removed from the cells along with the molten aluminum. In scrap melting operations, if the alloy being remelted contains magnesium, the dross will also contain some periclase (MgO) and spinel ($MgAl_2O_4$). Magnesium is preferentially oxidized to aluminum during remelting and so is present in higher fractions in the dross than in the original alloy. About 1%–5% of the charged aluminum winds up in the dross (Peel et al., 2011).

FIGURE 13.1 Nonsalt (*white*) dross. (Photograph courtesy of Ralf Urbach.)

FIGURE 13.2 Salt (*black*) dross. (Photograph courtesy of Aluminum Recovery Technologies, Inc., Kendallville, IN.)

Figure 13.2 shows salt dross (also known as black dross), the skim produced by furnaces remelting scrap with the use of flux (Peterson, 2011). This usually occurs during the remelting of smaller-sized or high-magnesium scrap and is most likely to be produced in sidewell reverberatory furnaces. Fluxing practice in reverbs has been previously described. Its purpose is to reduce metal losses by breaking up the oxide skin, thereby releasing the metal trapped inside. As a result, black dross usually contains less than 20% aluminum metal and 30%–50% aluminum oxide; the rest is the fluxing salt (mostly sodium chloride and potassium chloride). About 300–700 kg of black dross is generated per tonne of aluminum generated in a secondary smelting furnace (Romberg, 2007).

A third solid waste produced by some aluminum melting furnace is a material called *salt cake*. This will be described later.

PROCESSING OPTIONS FOR DROSS

Hot Processing

Figure 13.3 presents a hierarchy of options for processing and treatment of dross. The first option is whether to allow the skimmed dross to cool and solidify or to process it immediately after skimming, while the entrapped aluminum is still molten. Hot processing has two advantages:

1. It allows the operation to retain the heat energy present in the dross when it was first skimmed. If the dross is allowed to solidify, additional energy will be required to reheat it for further processing.
2. It offers greater potential recovery of the metal in the dross, since reprocessing occurs before the metal has a chance to oxidize.

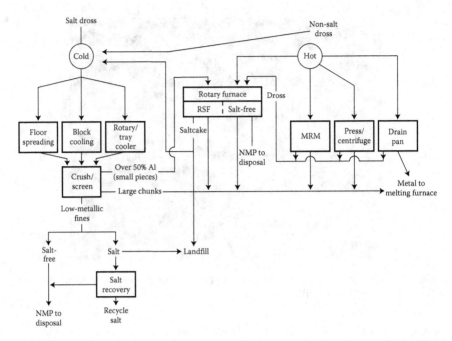

FIGURE 13.3 Process options for treatment of aluminum melting furnace dross. (From McMahon, J.P., Dross cooling and energy savings, in *Energy Conservation Workshop XI: Energy and the Environment in the 1990s*, Aluminum Association, Washington, DC, 1990. With permission.)

However, there are also disadvantages:

1. Equipment used for processing semimolten dross at 750°C–800°C is expensive to build and maintain.
2. Hot dross processing must be synchronized with production schedules in the melting furnaces.
3. Hot dross is subject to *thermiting*. Thermiting is the oxidation of the metal in the dross to form aluminum oxide or nitride (Peterson, 2011):

$$2Al + 1.5O_2 = Al_2O_3 \qquad (13.1)$$

$$Al + 0.5N_2 = AlN \qquad (13.2)$$

As the name suggests, thermiting is an exothermic (heat-generating) process. Uncontrolled, it can heat the dross to temperatures that can damage equipment. Furthermore, thermiting turns much of the metal in the dross into relatively worthless oxide. As a result, hot dross processing requires special techniques to minimize or eliminate thermiting.

Many traditional hot dross treatment processes involve heating the dross in a rotary salt furnace (RSF). As a result, the simplest way to retain the energy in hot dross is to simply charge it directly to the RSF without solidifying it first (Peel et al., 2011). The difficulty in doing this is synchronizing the skimming of the dross from the melting furnace and the operating cycle of the dross processing furnace. This can be solved either by treating the two as a single process under the control of one operator or by using special holders as a buffer, supplying an argon gas *blanket* to prevent thermiting (Spoel and Zebedee, 1996). Even so, coordinating activities between the multiple melting furnaces in a typical casthouse and a RSF often recommend a different practice.

A second approach to hot dross treatment involves agitating the dross to break the oxide skin surrounding the molten metal, allowing the metal to coalesce and ultimately be recovered. The metal reclaim machine, used extensively in east Asia, does this with an agitating impeller, recovering about 40% of the metal in the dross (Herbert, 2007; Okazaki et al., 1999). However, stirring tends to promote thermiting and turns the remaining dross into a fine dusty material from which further metal recovery is difficult. The more recent Drosrite process does this in a rotary furnace rotating at three revolutions per minute (Drouet et al., 2000). Drosrite uses an argon purge to solve the thermiting process. The AROS dross cooling unit also performs a similar function, as will be described later.

Two other hot dross processing technologies use pressure to squeeze the molten aluminum loose from the oxide skin. The first is the dross press. Figure 13.4 illustrates the basic principle behind dross pressing. Dross poured into the space between the press head and steel shell is squeezed by a hydraulic ram (Peel et al., 2011). Molten metal pours through the metallic drain in the center and into a sow mold. At the same time, the remaining dross is solidified. The solidified shell, a mixture of metal and oxide/salt, is removed for further processing. The amount of metal recovered is a function of press head design and the metal content of

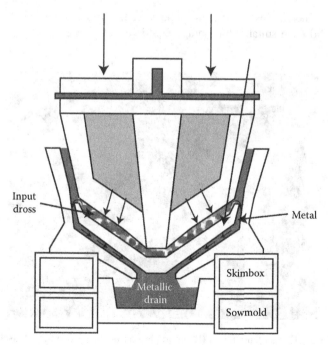

FIGURE 13.4 A typical dross press. (From Altek-MDY, Inc. With permission.)

the dross, but 60%–70% metal recovery is typical (Herbert, 2007). Optimal head design is an evolving concept, with extensive testing of new designs in recent years (Peel et al., 2011; Perry, 2000). Air cooling of the head is now standard, which improves dross cooling rates and further decreases metal losses by minimizing thermiting. The reduced thermiting in turn reduces aluminum carbide and nitride formation in the dross. Because of this, dross presses are currently the most popular choice for new dross cooling equipment. A less-expensive variation of this is the cooling head, which is simply set on top of the molten dross after it has been placed on a skim pan. The weight of the cooling head pushes 40%–60% of the metal in the dross through the holes in the skim pan. This simpler device is preferable for smaller operations.

The other use of pressure to squeeze molten aluminum loose from hot dross is centrifuging. In the Ecocent process (Ünlü and Drouet, 2002), hot dross is put in a *converter* and heated to 750°C (cold dross can also be charged) and stirred to *homogenize* it. The heated dross is then poured into a centrifuge (Figure 13.5), which forces the liquid metal through a screen against the side of the centrifuge. The centrifuge can generate either a ring of solidified metal for recharging to a furnace or molten metal that can be tapped from the bottom of the unit. Recovery of up to 90% of the input metal is claimed. Because the dross press and centrifuge are batch processes, they can be used any time a fresh *supply* of dross is skimmed from the furnace, eliminating the coordination problem associated with RSFs.

Dross Cooling Options

If the dross is not processed hot, it must be cooled. Dross cooling converts the material to a solid form suitable for storage or shipment and prevents the loss of metal

FIGURE 13.5 Dross centrifuge. (Photograph courtesy of Kärntner Maschinenfabrik, Villach, Austria.)

content by thermiting. Several technologies have been developed for dross cooling. Selecting the best means judging the alternatives against these criteria:

- Minimizes thermiting and loss of metal value
- Minimizes capital and operating costs
- Minimizes environmental impact
- Minimizes hazards to operating personnel
- Is logistically compatible with plant operations

The oldest approach to dross cooling is to simply spread the material on the floor and let it cool naturally (Herbert, 2007; Summer, 2009). While this has the lowest capital and operating cost of any option, it also results in the loss of about 70% of the metal value through thermiting. It also generates large quantities of oxide fume, which is hazardous both to plant personnel and to local air quality. In addition, the burning aluminum is a safety hazard to plant personnel. Some shops have tried spreading the dross on steel plates to encourage heat transfer and accelerate cooling. Pouring the dross into a water-cooled vibrating chute has also been tried, which offers some improvement. However, none of these open-air approaches eliminates the problems of floor spreading. As a result, they are gradually being abandoned.

Some use has been made of specially designed skimming boxes in which the dross is left to cool in a solid block. Thermiting can be limited by adding a salt blanket to cover the material and limit access to air (Figure 13.6). This process is slow and levies a financial penalty in the form of (a) the cost of providing the salt blanket or (b) metal losses from thermiting. An improved approach replaces the salt blanket with a cover that seals the dross container and prevents air access (Figure 13.7; Urbach, 2010). Improving the concept still further, the inert gas dross cooler (Figure 13.8) developed by Alcan provides an argon cover to exclude air (Herbert, 2007; Taylor and Gagnon, 1995), improving metal recovery over simple skimming boxes by 20%. However, dross cooled this way solidifies into a massive

FIGURE 13.6   Dross pan with salt cover. (Photograph courtesy of Ralf Urbach.)

FIGURE 13.7   Dross pan with sealed cover. (Photograph courtesy Ralf Urbach.)

block, requiring crushing to recover the metal. Dust generation is eliminated but subsequent processing costs are higher.

Introduced in the late 1970s, the rotary dross cooler was widely adopted in the 1980s for cooling and solidifying both salt and nonsalt drosses (Herbert, 2007; McMahon, 1990). Figure 13.9 illustrates the layout of a typical rotary dross cooler. The centerpiece of the device is a rotating drum fabricated from structural steel plate, with welded internal flights to help raise the dross as the drum rotates. Water is sprayed onto the outside of the top of the drum, flowing around the outside and collecting in a pool at the bottom. As it does, it extracts heat from the dross inside.

FIGURE 13.8   Inert gas dross cooler. (Photograph courtesy Ralf Urbach.)

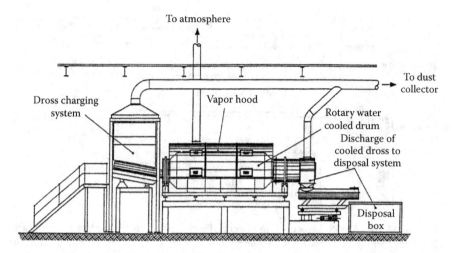

FIGURE 13.9 A rotary dross cooler. (From McMahon, J.P., Dross cooling and energy savings, in *Energy Conservation Workshop XI: Energy and the Environment in the 1990s*, Aluminum Association, Washington, DC, 1990. With permission.)

Some of the water vaporizes and leaves through the vapor hood at the top. Replacement water is added through a nozzle at the side. Although thermiting is reduced by the more rapid cooling in this unit, some does occur, and in addition the tumbling action crumbles the dross, generating more dust. As a result, dust collection hoods are located at both ends of the unit. Even so, the safety hazard presented by the dust and the presence of molten aluminum in the vicinity of a pool of water has decreased the popularity of rotary dross cooling in recent years.

Dross added to the cooler spends 5–15 min in the unit, during which its temperature is reduced below 50°C. The tumbling action reduces its size, making it easier to subsequently crush and screen. A trommel screen at the discharge end removes oversize material (+50 mm is typical). The oversize has a much higher metal content than the undersize, so it can be directly returned to the melting furnaces or further processed to remove more of the nonmetallic content. Overall aluminum recovery is 50%–60% (Herbert, 2007). Screen undersize is sent on for RSF melting or further solid processing.

The use of rotary dross cooling can improve metal recoveries by up to 50% over floor spreading, but the lack of an inert environment means that thermiting is still a concern, limiting overall recoveries to 40%–45% of the contained metal. Two approaches have been taken to solve the problem. Figure 13.10 illustrates the AROS dross cooler, introduced in the early 1980s (Roberts, 1990). The unit is sealed to minimize air intrusion and thus reduce thermiting. The water spray is replaced by a jacket placed around the central cooling drum. Cooled dross is discharged into the outside grinding drum, which turns with the cooling drum. Large lumps of metal in the discharge act as the *grinding balls* in this drum, knocking pieces of oxide and salt loose from the larger metal particles. A screen built into the grinding drum separates a coarse fraction (>8 mm), analyzing more than 90% metal. This material

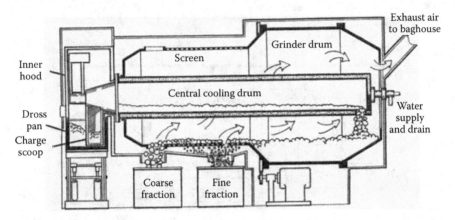

FIGURE 13.10 AROS dross cooling system. (From Roberts, R.P., In-plant processing of aluminum melting furnace dross, in *Energy Conservation Workshop XI: Energy and the Environment in the 1990s*, Aluminum Association, Washington, DC, 1990. With permission.)

can be directly remelted. Undersize from this screen is fed through a second screen built into the unit shell; the oversize from this *fine* screen (>0.2 mm) also analyzes 75%–80% metal. This product is sent for further upgrading. The undersize from the second screen, along with dust collected in the baghouse, is combined into a *dust* fraction. The metal content of the dust is less than 25% and can either be further processed for salt recovery or landfilled.

The second solution is to provide an inert environment (argon or nitrogen) to the rotary gas cooler. This improved metal recoveries by 15% over rotary dross cooling in air. Whether this improvement justifies the cost of the argon is uncertain.

Figure 13.11 illustrates the results from processing a single variety of dross, using different recovery methods (Herbert, 2007). Lighter colored bars at the bottom

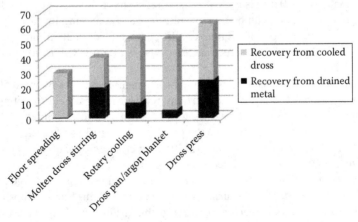

FIGURE 13.11 Metal recoveries from different dross processing techniques. (After Herbert, J., *Aluminium Times*, (3), 44, April/May 2007.)

indicate metal recovered by drainage from the processing unit; darker regions indicate aluminum recovery from subsequent rotary furnace processing. As might be expected, more expensive methods tend to produce the best results; whether the value of the additional metal recovered justifies the higher equipment cost depends upon a variety of factors.

COMMINUTION

With the exception of AROS units, dross coolers generate a product with a wide size range, from dust to chunks of several centimeters. As previously mentioned, the largest chunks have high enough metal content to remelt, while smaller pieces require upgrading to separate the metal from the oxide/salt content. Again, the dross processor has an option. Crushing the cooled dross will better *liberate* the metal from the attached oxides and salt, generating a product that is easier to directly remelt. However, both the capital and operating costs of crushing can be high, and the dust generated during crushing is an environmental concern. As a result, cooled nonsalt dross with over 50% metallic aluminum content is often fed directly to RSFs, while salt and lower-grade nonsalt drosses are crushed and upgraded to produce a concentrate with sufficient aluminum content.

Dross comminution is a multistage process, beginning with reduction of the largest chunks (>20 cm). This is typically performed using a shaft impactor or jaw crusher (Roth, 2011; Smith, 2008). The product is screened, and the oversize (>2.5 cm) material is returned to the melting furnace. The undersize from the screen is subsequently milled. Hammer mills have typically been used for this purpose, but cage mills may be more effective at crushing the nonmetallic particles without destroying the metal particles as well (Roth and Beevis, 1995). The mill product is again screened (300–500 μm); undersize is disposed of or processed to recover the salt, while oversize is used in RSFs to recover the metal.

Figure 13.12 shows the rotary tumbler (Roth, 2011). Originally developed in the 1970s for metal recovery from spent foundry sand, this device (currently in its third iteration) combines primary and secondary crushing, screening out large metal chunks as it processes dross.

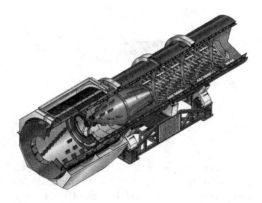

FIGURE 13.12   Rotary tumbler.

MELTING OPTIONS: ROTARY SALT FURNACE

Developed in the mid-1960s, the RSF is the primary method for recovering aluminum from dross. It is primarily used to treat high-grade nonsalt drosses and the concentrates produced by milling and screening salt and low-grade nonsalt drosses. There are several varieties of RSF, but the operating principle is similar.

Figure 13.13 illustrates a *typical* RSF. The furnace operates on a batch basis, with a total charge of 5–10 tons per batch (Gripenberg et al., 1995). The charge is a mixture of dross and/or dross concentrate, along with fluxing salt. The fluxing salt composition is a eutectic mixture of sodium chloride and potassium chloride, similar to that used in fluxed melting operations. As before, a flux with a few percent of an added fluoride salt (typically cryolite) helps to break the oxide skin around the molten aluminum as the contents are heated (Shell et al., 1995; also see Chapter 7). This allows the molten aluminum to coalesce, increasing recovery. The added salt also protects the metal underneath from oxidation by forming a molten blanket on top. The amount of salt added varies with the composition of the added dross, ranging from 50 to 500 kg of added salt per tonne of charged dross. In older RSFs, the goal is for the mass of salt to equal the mass of nonmetallic oxides in the dross. In newer furnaces, a lower ratio (0.2:1–0.4:1) is used. In either case, a higher-grade charge will use less salt than a low-grade charge.

In its typical configuration, energy is provided to the RSF by an air/fuel burner input through the side. Heat transfer from the flame to the charge is hindered by the molten salt layer on top, which has poor thermal conductivity. This problem is solved in the RSF by rotating the furnace. The refractories above the bath are heated by the flame. As the furnace rotates, the heated refractories move below the bath line

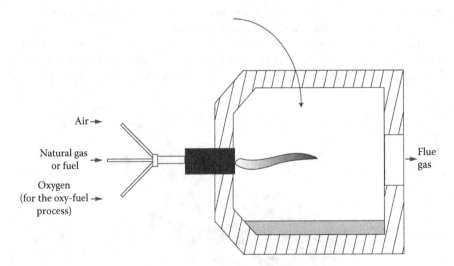

FIGURE 13.13 Cross section of a rotary salt furnace. (From Drouet, M.G. et al., A rotary arc furnace for aluminum dross processing, in *Third International Symposium on Recycling of Metals and Engineered Materials*, Queneau, P.B. and Peterson, R.D., Eds., TMS-AIME, Warrendale, PA, 1995. With permission.)

and transfer heat to the charge. RSFs can be fired with natural gas (more common) or fuel oil. The use of oxy-fuel burners has spread from scrap melting operations to RSFs, for the same reasons (Smith, 2008). The reduced nitrogen output from the furnace improves thermal efficiency, and the higher flame temperature increases the heat-transfer rate to the refractory. The use of regenerative burners has also spread from melting furnaces to RSFs. Fuel usage in an air/natural gas furnace is 3500–4000 kJ/kg of dross melted. When oxy-fuel burners and double-pass regenerators are installed, this figure can be reduced by as much as two-thirds. The use of oxy-fuel burners also allows the furnace to be sealed during operation (Peterson, 2011), allowing melting to take place in a reducing environment. This reduces melt loss and further reduces the need for flux.

Heating times are a function of the mass of charge and the heat-transfer rate. In older RSFs, tap-to-tap time can be 4 h or longer; in newer units, as little as 1. Nontilting RSFs generate two products. The first is molten aluminum, which is cast and sold as RSI. The second is a liquid nonmetallic product, which is cast into blocks known as *salt cake*. The salt cake (Figure 13.14) has a composition similar to that of salt dross and still contains 2%–10% metallic aluminum. Recoveries vary, but extracting 75% of the charged aluminum is not unusual. Salt-cake generation is 200–500 kg per metric ton of secondary aluminum produced (Tsakiridis, 2012).

The standard-model RSF has several drawbacks, the first of which is the production of liquid salt cake (wet processing). Tapping two liquid products from the furnace inevitably results in contamination, which can hurt metal purity or reduce recoveries. As a result, several dross processors now use a tiltable rotating barrel furnace (Gripenberg et al., 2002; Herbert, 2007; Smith, 2008). Furnaces like this (Figure 13.15) allow the molten aluminum to be decanted from the vessel first, so that a second liquid product is no longer needed. As a result, a *dry* dross smelting process can be used, in which the salt usage is reduced by half and a solid salt cake (5%–12% Al, 25%–50% flux, remainder oxides) is produced. This reduces materials

FIGURE 13.14   Salt cake. (Photograph courtesy of Ralf Urbach.)

FIGURE 13.15 Tiltable rotating furnace.

and energy costs and makes product removal easier. Charging clean scrap along with dross reduces the salt requirement still further.

However, even the dry process requires salt additions. This results in salt vapor emissions from the furnace, along with the production of salt cake. The salt cake is expensive to process, as will be discussed later, and cannot be landfilled in many countries. As a result, considerable efforts have been made over the past 20 years to develop salt-free methods for recovering aluminum from nonsalt dross.

SALT-FREE PROCESSES

Initial attempts to operate rotary barrel furnaces without flux were unsuccessful, largely because the lack of a salt cover allowed thermiting to occur. The exothermic reaction caused rapid heating of the charge, which increased the thermiting still further (Peterson, 2011). As a result, metal recoveries were poor (55%–60%) and frequent relining was required. In response to this, development efforts produced four new salt-free melting technologies (Simonian, 2001).

The simplest of the four is ALUREC (Figure 13.16), developed by AGA in partnership with Hoogovens Aluminium and MAN GHH (Gripenberg et al., 1995, 1997). ALUREC uses an oxy-fuel burner as an energy source, with a sealed furnace door to keep air out. Replacing air with pure oxygen eliminates nitrogen from the furnace environment. This reduces the flame size, meaning less direct heating of the dross. It also increases flame temperature, improving heat transfer to the walls and increasing the heating rate of the charge. As with other salt-free processes, ALUREC relies on the mixing generated by furnace rotation to break the oxide skin on the metal in the dross, allowing it to coalesce. Operation of a pilot plant at Hoogovens yielded recovery of up to 90% of the metal in the dross, with fuel usage of less than 1300 kJ/kg of dross processed. The solid salt cake recovered from the furnace is half of that resulting from RSF processing. Reducing the amount of flux to purchase and the amount of salt cake to process gave ALUREC a 45% cost advantage over standard

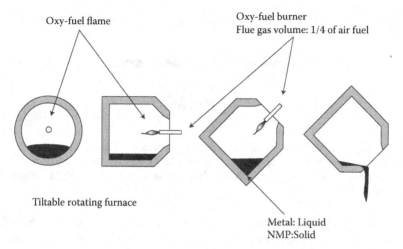

FIGURE 13.16 Schematic of ALUREC dross processing technology. (From Gripenberg, H. et al., ALUREC—A new salt-free process, in *Third International Symposium on Recycling of Metals and Engineered Materials*, Queneau, P.B. and Peterson, R.D., Eds., TMS-AIME, Warrendale, PA, 1995. With permission.)

RSF processing. However, ALUREC technology has not become widespread due to high capital costs and lower recoveries.

The most widely publicized of the four processes is the plasma dross treatment technology introduced in the early 1990s by Alcan (Ünlü and Drouet, 2002). Figure 13.17 shows a simplified view of a nontransferred-arc plasma torch. A gas (air or nitrogen) is passed through the gap between the two copper-alloy electrodes.

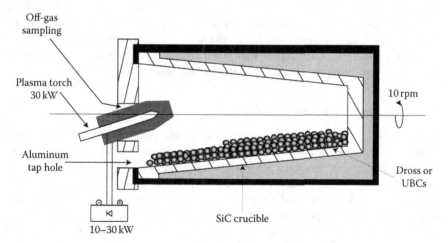

FIGURE 13.17 Nontransferred-arc plasma dross treatment furnace. (From Kassabji, F. and Weber, J.C., Aluminum scrap processing in a rotary plasma furnace, in *Extraction and Processing for the Treatment and Minimization of Wastes*, Hager, J. et al., Eds., TMS-AIME, Warrendale, PA, 1993. With permission.)

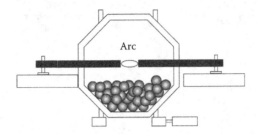

FIGURE 13.18 DROSCAR transferred-arc plasma furnace. (From Kassabji, F. and Weber, J.C., Aluminum scrap processing in a rotary plasma furnace, in *Extraction and Processing for the Treatment and Minimization of Wastes*, Hager, J. et al., Eds., TMS-AIME, Warrendale, PA, 1993. With permission.)

Applying high-voltage power to the torch generates an electric arc between the electrodes that heats the gas to plasma temperatures (>5000°C). Gases this hot give off nearly all of their available heat by radiation, resulting in much higher thermal efficiency than that of an RSF. As a result, gas usage is minimal, resulting in much lower off-gas volumes than RSF. This reduces oxidation and results in metal recoveries of over 95%. Again, salt-free processing yields the same advantages as previously described. However, plasma dross processing has also not succeeded commercially. High electricity prices and high capital costs are the likely causes.

Figure 13.18 shows a DROSCAR transferred-arc plasma furnace, using graphite electrodes (Drouet et al., 1994, 1995; Meunier et al., 1999). Again, the application of high voltage causes an arc to be struck between the electrodes, generating high temperatures and heat transfer by radiation to the walls and directly to the charge. As with the other salt-free processes, the DROSCAR process seals the furnace to keep air out. As a result, gas volumes are even lower than for the nontransferred-arc plasma (3 m³/ton of processed dross vs. 30 for the Alcan furnace—RSF processing generates 300 m³/ton). An argon environment is provided to stabilize the arc; this also reduces thermiting and limits the formation of aluminum nitride. This results in metal recoveries similar to the other salt-free processes. Again, high electricity costs can be a concern, and lumpy dross pieces can break the electrodes as they tumble during furnace rotation.

SALT CAKE AND SALT DROSS PROCESSING

As previously pointed out, it is common practice to recover much of the aluminum from salt dross by crushing and concentration. However, this still leaves a *salt-cake* residue, consisting of flux salts (25%–60%), aluminum nitride, metallic aluminum (4%–10%), and various oxides (Tsakiridis, 2012; Urbach, 2011). The salt content is higher in salt cake from fixed rotary furnaces and lower from that in tiltable furnaces. Land-based disposal of this material is still economically feasible in the United States but is either banned or too expensive elsewhere. Because of this, further treatment is required. The goals of this further processing include the following:

- Minimizing or eliminating the residue to be discarded
- Generating a nonhazardous residue that can be discarded if necessary

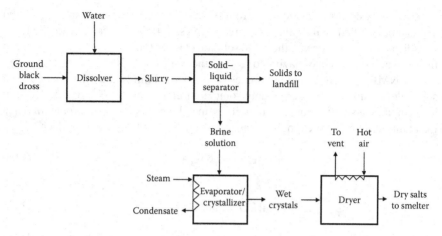

FIGURE 13.19 Standard processing flowsheet for salt cake. (From Sheth, A.C. et al., *JOM*, 48(8), 1996. With permission.)

- Recovering the salt content in the feed
- Recovering the metallic aluminum in the feed
- Reduced cost and complexity for the process
- Minimizing environmental impact of the process

Figure 13.19 illustrates a basic process used for most salt-cake treatment (Sheth et al., 1996; Tsikiridis, 2012). The feed has been crushed and screened as previously described to remove most of the aluminum metal. Since the water will have to be removed later, only enough is added to produce a brine with about 22%–25% dissolved salt content. The salts (primarily NaCl and KCl) generate heat when they dissolve, raising brine temperatures as high as 60°C. The solid–liquid separator is usually a two-stage process, beginning with a centrifuge, which separates out most of the NMP (Non-metallic product, i.e., the oxide and metal content). The liquid leaving the centrifuge is then passed into a clarifier, which generates a sludge containing the rest of the NMP. This is washed and filtered, resulting in a low-salt NMP that can be landfilled or possibly used for other purposes. This purifies the brine and also removes solids such as calcium sulfate and magnesium hydroxide. These can cause scaling in the crystallizer if not removed.

The brine generated by dissolution is then processed to remove the water. This is usually done with an evaporator crystallizer. These crystallizers have traditionally been multi-effect forced circulation units, but mechanical vapor recompression systems have higher efficiencies and have become a preferred option. More recent development by Engitec has promoted the use of flash evaporation crystallization (Sheth et al., 1996). This reduces energy requirements, at the cost of additional process complexity. The result of crystallization is wet salt crystals that are subsequently air dried and reused as flux. The potassium chloride in melting flux is preferentially vaporized during melting, so the salt recovered from the brine has a higher NaCl/KCl ratio than the salt used as flux. Purchased KCl is added to bring the ratio back to the desired level.

The energy costs of salt recovery from salt-cake processing can be substantial and may not be justified by the value of the recovered salt. If the salt-cake treatment facility is located near the ocean, the brine solution can be simply diluted and returned to the sea, saving the cost of having to recover the salt (Smith, 2008).

The NMP recovered from the process consists of alumina and other oxides, with some aluminum nitride and possibly some aluminum carbide (Tsakiridis, 2012). It also contains a small amount of remaining metal, and this poses a problem. When the aluminum nitride is exposed to moisture, alumina and oxide are produced:

$$2AlN + 3H_2O = 2NH_3 + Al(OH)_3 \tag{13.3}$$

$$2AlN + 4H_2O = Al(OH)_3 + NH_4OH \tag{13.4}$$

The alkaline environment strips the oxide coating from the aluminum particles in the NMP, and this allows moisture to react with the metal and generate hydrogen:

$$3H_2O + 2Al = Al_2O_3 + 3H_2 \text{ (+ heat)} \tag{13.5}$$

The presence of hydrogen and heat is a fire and explosion hazard, and this makes the disposal of NMP a problem. Aluminum carbide in the NMP engages in similar reactions, generating methane and possibly other hazardous by-products. As a result, disposal of this material is still potentially hazardous, and several of its potential uses—cement production, use in the production of calcium aluminate—are impacted as well.

RECOMMENDED READING

Herbert, J., The art of dross management, *Aluminium Times*, (3), 44, April/May 2007.
Peterson, R.D., A historical perspective on dross processing, *Mater. Sci. Forum*, 693, 13, 2011.

REFERENCES

Drouet, M.G., Handfield, M., Meunier, J., and Laflamme, C.B., Dross treatment in a rotary arc furnace with graphite electrodes, *JOM*, 46(5), 26, May 1994.
Drouet, M.G., LeRoy, R.L., and Tsantrizos, P.G., DROSRITE salt-free processing of hot aluminum dross, in *Fourth International Symposium on Recycling of Metals and Engineered Materials*, Stewart, D.L., Stephens, R., and Daley, J.C., Eds., TMS-AIME, Warrendale, PA, 2000, p. 1135.
Drouet, M.G., Meunier, J., Laflamme, C.B., and Handfield, M.D., A rotary arc furnace for aluminum dross processing, in *Third International Symposium on Recycling of Metals and Engineered Materials*, Queneau, P.B. and Peterson, R.D., Eds., TMS-AIME, Warrendale, PA, 1995, p. 803.
Gripenberg, H., Gräb, H., Fleschm, G., and Mullerthann, M., ALUREC—A new salt-free process, in *Third International Symposium on Recycling of Metals and Engineered Materials*, Queneau, P.B. and Peterson, R.D., Eds., TMS-AIME, Warrendale, PA, 1995, p. 819.
Gripenberg, H., Lodin, J., Falk, O., and Niedermair, F., New tools for melting of secondary aluminum in rotary furnaces, *Aluminium*, 78(9), 642, 2002.

Gripenberg, H., Müllerthan, M. and Jäger, N., Aluminium dross and waste recycling with ALUREC, in *3rd ASM Int. Conf. Recycl. Met.*, ASM International Europe, Brussels, Belgium, 1997, p. 421.

Herbert, J., The art of dross management, *Aluminium Timex*, (3), 44, April/May 2007.

Kassabji, F. and Weber, J.C., Aluminum scrap processing in a rotary plasma furnace, in *Extraction and Processing for Treatment and Minimization of Wastes*, Hager, J. et al., Eds., TMS-AIME, Warrendale, PA, 1993, p. 687.

van Linden, J.H.L., Melt loss and dross treatment, in *Fifth Australasian-Asian Pacific Conference on Aluminum Cast House Technology*, Nilmani, M., Whiteley, P., and Grandfield, J., Eds., TMS-AIME, Warrendale, PA, 1997, p. 71.

McMahon, J.P., Dross cooling and energy savings, in *Energy Conservation Workshop XI: Energy and the Environment in the 1990s*, Aluminum Association, Washington, DC, 1990, p. 165.

Meunier, J., Laflamme, C.B., and Biscarro, A., DROSCAR-RESIMIX: an efficient and environmentally sound process to recover aluminium from dross, in *REWAS '99*, Vol. II, Gaballah, I., Hager, J., and Solozabal, R., Eds., TMS-AIME, Warrendale, PA, 1999, p. 985.

Okazaki, H., Takai, M., Hayashi, N., Uehard, T., and Ohzono, T., Effect of atmosphere on metal recovery from aluminum dross, in *REWAS '99: Global Symposium on Recycling, Waste Treatment and Clean Technology, Vol. II*, Gaballah, I., Hager, J., and Solozabal, R., Eds., TMS-AIME, Warrendale, PA, 1999, p. 975.

Peel, A., Herbert, J., Connaughton, R., and Cotton, C., Preserving metal units utilizing the latest generation of aluminum dross pans, *Mater. Sci. Forum*, 693, 33, 2011.

Perry, O.H., The development of the modern dross press, in *Light Metals 2000*, Peterson, R.D., Ed., TMS-AIME, Warrendale, PA, 2000, p. 675.

Peterson, R.D., A historical perspective on dross processing, *Mater. Sci. Forum*, 693, 13, 2011.

Prakash, M., Cleary, P.W., and Taylor, J.A., SPH modeling of the effect of crucible tipping rate on oxide formation, *Mater. Sci. Forum*, 693, 54, 2011.

Roberts, R.P., In-plant processing of aluminum melting furnace dross, in *Energy Conservation Workshop XI: Energy and the Environment in the 1990s*, Aluminum Association, Washington, DC, 1990, p. 143.

Romberg, R., Aluminium salt slag recycling – Optimisation within the ALS Process, presented at *9th Int. Alum. Recycl. Congr.*, OEA, Düsseldorf, 26 February, 2007.

Roth, D.J., The approach to zero waste from smelter and secondary dross processing, *Mater. Sci. Forum*, 693, 24, 2011.

Roth, D.J., and Beevis, A.R., maximizing the aluminum recovered from your dross and elimination of any waste products in dross recycling, in *Light Metals 1995*, Evans, J., Ed., TMS-AIME, Warrendale, PA, 1995, p. 815.

Shell, D.J., Nilmani, M., Fox, M.H., and Rankin, W.J., Aluminium dross treatment using salt fluxes, in *Fourth Australasian-Asian Pacific Conference on Aluminum Cast House Technology*, Nilmani, M., Ed., TMS-AIME, Warrendale, PA, 1995, p. 133.

Sheth, A.C., Parks, K.D., and Parthasarathy, S., Recycling salt-cake slag using a resin-based option, *JOM*, 48(8), 32, August 1996.

Simonian, G., Comparison of rotary salt and non salt dross processing technologies, in *Seventh Australasian-Asian Pacific Conference on Aluminum Cast House Technology*, Whiteley, P.R., Ed., TMS-AIME, Warrendale, PA, 2001, p. 115.

Smith, T., Alumox and Reox: The treatment of dross and salt cake, *Aluminium International Today*, 20(2), 18, 2008.

Spoel, H. and Zebedee, W.A., The hot aluminum dross recycling (HDR) system, in *Light Metals 1996*, Hale, W., Ed., TMS-AIME, Warrendale, PA, 1996, p. 1247.

Summer, H., New methods for preparing salt-containing and salt-free dross from the primary or secondary aluminium smelting sector, *Aluminium*, 85(3), 40, 2009.

Taylor, M.B. and Gagnon, D., The inert gas dross cooler (IGDC), in *Light Metals 1995*, Evans, J., Ed., TMS-AIME, Warrendale, PA, 1995, p. 819.

Tsakiridis, P.E., Aluminium salt slag characterization and utilization—A review, *J. Hazard. Mater.*, 217–218, 1, 2012.v

Ünlü, N. and Drouet, M.G., Comparison of salt-free aluminum dross treatment processes, *Resour. Conserv. Recycl.*, 36, 61, 2002.

Urbach, R., Where are we now in the field of treatment of dross and salt cake from aluminium recycling, *International Aluminium Recycling Workshop*, 11 June 2013, Trondheim, Norway, 2010.

14 Safety and Environmental Considerations

In addition to its economic advantages, aluminum recycling has desirable environmental benefits. Chapter 1 pointed out that production of aluminum from recycled sources reduces CO_2 emissions to the atmosphere by 90% compared with primary metal, in addition to reducing fluoride and SO_2 emissions. The production of secondary metal reduces energy usage by over 90% compared with primary metal and reduces the amount of waste generated by 90%.

However, the technology of aluminum recycling presents potential hazards of its own to both the work force and the general public. As the tolerance of the public for these hazards decreases, the need to improve the performance of recycling operations grows. This last chapter will explore the unique hazards of aluminum recycling and describe the measures taken in response. The result of these measures will be an industry that moves into the future with confidence in its value to society.

COLLECTION AND BENEFICIATION

Chapter 3 described the two basic types of metal scrap, new and old. New scrap is that obtained directly from manufacturing operations, while old scrap is obtained from discarded manufactured items. The fraction of new scrap that is recycled is higher than that of old scrap; however, the fraction of old scrap being recycled is increasing, due to the limited amount of new scrap available and the cost of producing primary aluminum in some parts of the world (Europe in particular).

Statistics are difficult to find on the relative hazards of employment in scrap collection and beneficiation. The US Department of Labor (BLS, 2012) currently maintains industrial injury data for two industry classifications: "recyclable material merchant wholesalers" and "materials recovery facilities." The first classification includes wholesalers handling a wide variety of recyclables, including boxes, glass, rubber, and fur cuttings as well as metallic scrap (United States Census Bureau, 2012). The second classification includes both public facilities (MRFs, described in Chapter 6) and private sorting facilities. Table 14.1 compares injury rate data for workers in these classifications with those for private industry as a whole for three recent years (the only ones for which data are available). The results show a higher rate of both overall injuries and serious injuries in recycling-related industries. Statistics compiled by the British Health and Safety Executive (HSE, 2012) show that major injuries in the waste management and recycling industry happen at over four times the rate as industry as a whole. Neitzel et al. (2013) have shown high risks for a number of specific hazards at scrap yards, in particular noise, repetitive motion, and laceration/abrasion hazards.

TABLE 14.1

Injury Rates in US Scrap Collecting and Processing

Year	Total Recordable Injuries (per 100 Workers)			Serious Injuries (per 100 Workers)		
	RM	MR	PI	RM	MR	PI
2007	5.1		4.2	3.1		2.1
2008	7.3	7.0	3.9	3.8	4.0	2.0
2010	5.6	3.4	3.5	3.2	1.8	1.8

Source: BLS, United States Department of Labor, Bureau of Labor Statistics, Injuries, illnesses, and fatalities, http://www.bls.gov/iif/oshsum.htm, 2012.

Note: RM, recyclable material merchant wholesalers; MR, materials recovery facilities; PI, private industry.

In general, the collection of new scrap presents fewer hazards and environmental concerns than that of old scrap. Because new scrap is produced in a limited number of locations, it is easier to collect and rarely requires separation from other materials. New scrap is purer and so requires less processing; impurities that might pose workplace or environmental hazards are less likely to be present. Finally, new scrap is a *known* and contains fewer surprises. The increasing number of captive recycling operations that accept scrap from one industrial facility and return secondary ingot directly to that facility makes processing new scrap even less hazardous.

Old scrap is more difficult to process safely and presents several hazards. These include the following:

- *Hazardous chemicals*: All sorts of chemicals are put in aluminum containers, and recovering these chemicals without exposure of the workforce or the general public can be difficult. Larger items such as discarded automobiles and appliances are especially likely to contain fluids (PCB-containing oils, refrigerants) requiring special handling (Bertram et al., 2007; Dougherty, 2000).
- *Labor usage*: Because old aluminum scrap has to be separated from other materials, recycling is often labor intensive. The use of hand sorting was discussed in Chapter 6; disassembly of large items such as ships or aircraft is also done largely by hand. This exposes workers to the hazards (heat, noise, dust) normally associated with this type of work, in addition to the unique hazards of demolition work (falling objects, asbestos in ship breaking). Cutting equipment such as shears or saws can be especially hazardous. The US Occupational Safety and Health Administration (OSHA, 2008) lists *contact with an object or piece of equipment* and *overextension* as two of the most common causes of lost-time injuries in scrap metal processing.

- *Explosive material*: Mention was briefly made in Chapter 6 of the problems posed by aerosol cans in scrap. Shredded under pressure, these cans can explode (BAMA, 2002), and the contents can make the explosion worse if they are flammable. Military scrap can also contain explosive material if not properly decommissioned (Jacoby, 2000; Minter, 2006). Items as small as a butane lighter can become an explosion hazard if not spotted (Bertram et al., 2007).
- *Radioactive material*: Although no radioactive aluminum alloys are commercially produced, mixed scrap can contain other radioactive sources that have not been properly disposed of. These radiation sources can contaminate an entire recycling facility if not caught and isolated quickly (Bertram et al., 2007).
- *Location issues*: Many scrap yards were located on the edge of town when they first began operating several decades ago. These yards now find themselves in the middle of an expanded urban area. The residents of adjacent neighborhoods find scrap collectors and processors an undesirable neighbor, and the heavy equipment and hazardous chemicals represent a threat to those living near as well as working at the yard (*Dallas Morning News*, 2011; Malloy, 2011; Woodin, 2012).

The response to these hazards varies considerably with location and the type of scrap upgrading technology used. However, several common strategies are available. These include the following:

- *Automation*: Although hand picking is still widely practiced in Third World countries, the hazards that it poses to the workforce are increasingly unacceptable in more advanced societies. As a result, minerals processing technology (optical sorting, specific gravity separation, eddy current devices) is replacing hand pickers in many locations.
- *Improved equipment*: The equipment used by scrap processors is much safer to operate than it was a generation ago. Shears have guards to protect operators; hammer mills have exhaust filters to remove dust; explosion-proof shredders are designed to withstand the impact of an exploding aerosol can or flammable gas mixture. A current focus is improving baler safety, using devices that sense when an operator is in the main chamber and prevent the ram from activating (OSHA, 2008).
- *Collection by type*: The hazards of processing old scrap differ according to the product being recycled. UBCs have lacquer coatings not found on automotive scrap; discarded automobiles contain motor oil and fluids not found in UBCs. Separating old scrap by product type makes it easier to deal with these hazards. Automobiles and UBCs are already processed separately in dedicated facilities in many locations, as are aircraft and old electrical cable (see Chapter 6). Efforts are being made to separate white goods into a distinct processing stream as well (GIZ, 2011; Ludwig, 1996).

- *Regulation of scrap and sorting facilities*: Increasingly, governments have become involved in directing the movement of scrap and the operation of facilities for processing (Los Angeles County, 2011). The goals of the regulatory process include eliminating the uncontrolled dumping of hazardous chemicals recovered from scrap, protection of the general public from noise and discharges from the facility, and ensuring worker safety (Headland Multimedia, 2012; Missouri Department of Natural Resources, 2004; OSHA, 2008). In some cases, scrap yards in populated areas have been forced to close or curtail operations. The regulatory process often goes hand-in-hand with the writing of standards (see the following).
- *Development of standard operating practices*: The development of safe work practice in scrap yards and MRFs until recently followed standards used at construction sites (Glenn, 1998). These are of limited value, since the equipment used for tearing items apart is different from that used for putting things together. However, two US initiatives have recently been announced to develop more industry-specific standards. The first is the publication of ANSI (American National Standards Institute) Z245.41, which applies to MRF operations. More recently, the industry group ISRI has developed RIOS (Recycling Industries Operating Standard), which standardizes responses to environmental issues (Wagger, 2010). OSHA (2008) lists numerous operating standards applicable to different unit processes in scrap collection, sorting, and processing.

THERMAL PROCESSING AND MELTING

Facilities that remelt aluminum scrap receive raw material both from scrap processors that upgrade old scrap and directly from producers of new scrap. As a result, the environmental and safety hazards faced by processors are also present at remelting facilities. Additional hazards include the following (Bertram et al., 2007; Jacoby, 2000):

- *Water*: There are numerous potential sources of water in melting and casting facilities (Ekenes, 2007); however, the most serious accidents usually arise from charging wet scrap (Epstein, 2007). Moisture comes with scrap either as trapped liquid, as snow and ice accumulated during storage (Pluchon et al., 2011), or as rainwater collected during transportation. In any case, it is a significant safety hazard. If water is charged to a melting furnace along with the scrap, the resulting steam explosion can injure or kill plant personnel and damage equipment. The problem is serious enough that separate statistics are specifically kept for steam explosions.
- *Rust*: Rust is hydrated iron oxide (Fe_2O_3). It represents a safety hazard because of its reactivity with molten aluminum:

$$Fe_2O_3 + 2Al\ (l) = 2Fe\ (l) + Al_2O_3 \qquad (14.1)$$

This *thermite* reaction is highly exothermic, to the point of being explosive (Ekenes, 2007; Pierce et al., 2005). The excess heat caused by the reaction can also lead to more oxidation of the molten metal, further increasing melt loss.

- *High-temperature toxics*: These are chemicals that pose no hazard during ambient-temperature processing and handling but give off toxic fumes when heated (Bertram et al., 2007). The organic coatings on UBCs are a good example of this; the selenium and arsenic coating of photoreceptor tubes is another.
- *Fines and dust*: These can be generated during shredding or wire chopping and collect on the floor of storage bins or in the dust collection system (Bertram et al., 2007; Léon, 2005; Pierce et al., 2005). The resulting explosions can be catastrophic (CSHIB, 2005).
- *High-temperature explosives*: The reaction of nitrate chemicals with molten aluminum when charged with a scrap load is a notorious safety problem in remelting operations. The most common source is ammonium nitrate from discarded fertilizer (Bertram et al., 2007; Pierce et al., 2005). The ammonium phosphate in discarded fire extinguishers also reacts violently with molten aluminum. Oxidized material has on occasion been mistaken for flux, with deadly results (Caniglia, 2004). A growing problem is that of air bags and seatbelt pretensioners from automobiles, which contain explosive sodium azide. Most of these are detonated in the shredder, but some occasionally make it through intact.
- *Chlorine gas and/or flux*: Mention was made in Chapter 12 of the desire by the aluminum industry to reduce and if possible replace the use of chlorine gas in refining operations (DeYoung and Levesque, 2005). One important reason for this is the emissions that result from the use of chlorine, which include both Cl_2 and HCl vapors as well as particulate matter. These emissions occur at levels well below regulatory limits, but are still a source of concern. The salt flux used as a replacement for chlorine in rotary and some reverberatory furnaces vaporizes to a small extent during melting, generating chloride fumes (Chesonis et al., 2009). Dust given off by salt slag also presents a workplace hazard. More recently, the hygroscopic nature of salt accumulated on workplace tools has been identified as a path by which water can be introduced to melting furnaces (Williams et al., 2005) and thus a cause of steam explosions.

Figures 14.1 and 14.2 show the relative injury and serious rates of the *secondary smelting and alloying of aluminum* industry compiled by the US Department of Labor (BLS, 2012). Although they have in general declined with time, the values are significantly higher than for private industry as a whole. Much of the reason lies in the persistent incident of molten metal explosions, which have caused numerous injuries and fatalities. Figure 14.3 shows historical data collected by the Aluminum Association (United States) on the global number of *incidents* involving molten aluminum explosions (Epstein, 2011). The most numerous of

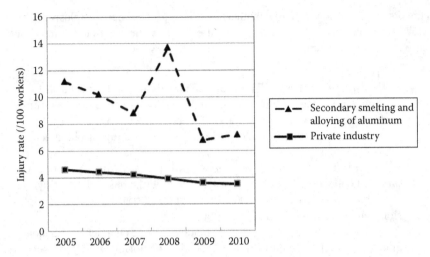

FIGURE 14.1 Injury rates in the secondary aluminum smelting industry and US primary industry as a whole. (From BLS, United States Department of Labor, Bureau of Labor Statistics, Injuries, illnesses, and fatalities, http://www.bls.gov/iif/oshsum.htm, 2012.)

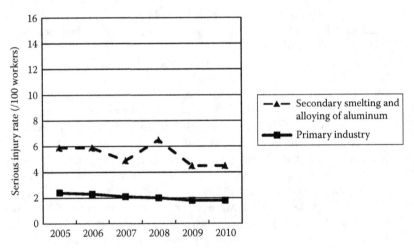

FIGURE 14.2 Serious injury rates in the secondary aluminum smelting industry and US primary industry as a whole. (From BLS, United States Department of Labor, Bureau of Labor Statistics, Injuries, illnesses, and fatalities, http://www.bls.gov/iif/oshsum.htm, 2012.)

the three types of explosion are Force 1 explosions, sometimes known as *pops* or *steam explosions* (Epstein, 2007). Figure 14.4 shows a typical example; these occur when water trapped under molten metal expands and throws a small amount of metal (<4 kg) a short distance (<5 m). Less numerous are Force 2 explosions, when the steam is trapped by something more resistant than just molten metal

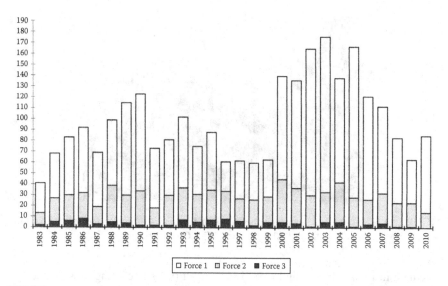

FIGURE 14.3 Aluminum Association data for global molten aluminum *incidents*. (From Epstein, S.G., Annual summary report on molten metal incidents, http://www.aluminum.org/AM/CM/ContentDisplay.cfm?ContentFileID=65185&FusePreview=Yes, 2011. With permission.)

(e.g., unmelted solid), resulting in an explosion that can throw molten metal up to 15 m or more. These are capable of doing serious damage to both plant equipment and personnel. Force 3 explosions, which result in the type of damage seen in Figure 14.5, are rare (for the first time, none were reported in 2010). These are often *assisted* by chemical reactions, such as exploding ammonium nitrate or the reaction of molten aluminum with water to generate hydrogen (Lowery and Roberts, 2010):

$$2Al + 3H_2O = Al_2O_3 + 3H_2 \qquad (14.2)$$

FIGURE 14.4 A typical Force 1 incident.

FIGURE 14.5 Force 3 explosion at the Binzhou Weiqiao Aluminum Company in China, August 2007. (Photo courtesy of Alex Lowery, Wise Chemical.)

The data in Figure 14.4 include explosions of all types, including those associated with casting, metal transfer, and melting. About 15% resulted in a serious injury or fatality (Pluchon et al., 2011). Melting incidents represented about 40% of the total, but almost two-thirds of the fatalities. As a result, the elimination of these incidents is a priority for the industry.

Perhaps the most significant concern resulting from the use of chlorides is their reaction with organic vapors given off by decomposing coatings and plastics. The reaction generates several persistent organic pollutants (POPs), most notably PCDDs (polychlorinated dibenzo-*p*-dioxins) and PCDFs (polychlorinated dibenzo furans). These compounds are found in both the off-gas from melting furnaces and the fly ash captured in the dust cleaning system (Pitea et al., 2008). Total emissions and emission rates vary considerably among secondary aluminum producers, due to variations in the type of scrap used, melting practice, and off-gas treatment method (Ba et al., 2009; Lee et al., 2009). The types of PCDD/F found in the off-gas and fly ash are largely the same, with furans predominating over dioxins (Lenoir et al., 2012; Li et al., 2007). The most common congeners (forms of PCDD and PCDF) are a matter of some disagreement. Polychlorinated naphthalenes (PCNs) are also found in the off-gas and fly ash from secondary aluminum smelting (Ba et al., 2010).

The high ratio of furan to dioxin congeners in secondary aluminum furnace off-gas suggests that the primary formation mechanism for these compounds is *de novo* synthesis (Lee et al., 2005; Lenoir et al., 2012; Li et al., 2007; Pitea et al., 2008). This is a reaction that occurs in the off-gas as it leaves the furnace, using the solid carbon in the fly ash from the furnace as a carbon source (Merz, 2004). Molecular oxygen in the off-gas stream is also a requirement. The source of the needed chlorine is less certain; changing concentrations of Cl_2 and HCl have little impact on the formation rate, so the presence of a metal catalyst such as copper may play an important role.

Table 14.2, from the 2001 review by Nakamura et al., compares dioxin emissions from various industries in Europe and the United States. (A g-TEQ is the equivalent mass of toxic dioxin emitted per year; 1 g-TEQ is roughly equal to 61 g of total dioxin.) Ba et al., show similar results from a survey in 2009. The numbers are

TABLE 14.2

Estimated Dioxin Emissions from Various Countries by Source

Source Category	Emissions/Year (g-TEQ)				
	United States (1995)	Japan (1999)	United Kingdom (1997/1998)	Germany (1994/1995)	38 European Countries (1990s)
Waste incineration (total)				32	2660
Municipal waste incineration	1100	1350	710–2600		
Industrial/hazardous waste incineration	5.7	690			
Ferrous metal industry (total)				181	1960
Sintering plants	25.8	101.3	25–30	158	1650
Electric arc furnace		141.5	59	5	287
Nonferrous metal industry (total)					1610
Primary copper	0.5	0.46	24	26.8	1500
Secondary copper	541	0.05			
Primary lead		0.04			18.0
Secondary lead	1.63	0.45	95–220		
Primary zinc		0.14			30.0
Secondary zinc		18.4		41.8	
Aluminum industry (total)		14.0		49.7	37.3
Primary aluminum			0.082		
Secondary aluminum	0.23		29–320	22.9	

Source: Nakamura, T. et al., *Metall. Rev. MMIJ*, 17, 93, 2001. With permission.

subject to considerable uncertainty, but the much larger values for the European secondary aluminum industry may reflect in part the wider use of rotary furnaces, which require fluxing (Sweetman et al., 2004). Dioxin generation is facilitated in the remelting of aluminum scrap because of (a) the melt temperature of 700°C–800°C, which is below the temperature at which optimal decomposition of dioxins occurs and (b) the input raw material, which often includes chlorine- and bromine-containing plastics. Electronic scrap is a particular concern (Sinkkonen et al., 2004).

RESPONSES TO HAZARDS

The responses to these hazards fall into two classes. The first has to do with scrap acquisition and storage and the second is changes in charging and melting practice.

Scrap acquisition and storage: Preventing steam explosions is the most important task in the receiving and handling of scrap aluminum (or any metal scrap, for that matter). Visual inspection of incoming scrap can determine if a load is especially wet, but measures taken after receiving are more important (Pluchon et al., 2011). Storing scrap in a covered location is recommended, as is a FIFO (First In, First Out) policy for charging. Although baled scrap is more convenient to charge, shredding the bales is increasingly required to uncover surprises and give undesired fluids a chance to drain away (Bano Shredders, 2011; Jacoby, 2000). Storage temperatures should be watched carefully, as sudden changes in temperature can cause condensation of atmospheric moisture onto the scrap surface to occur.

Inspection of incoming scrap is vital for removing other hazards, and training personnel to recognize hazards is an important part of a safety program (Bertram et al., 2007). Many facilities now supplement visual inspection with a radiation detection station, since radioactive material is the hardest to spot and the most damaging if not caught. Magnetic separators can remove most of the ferrous material (and the rust) from incoming scrap, and this is commonly practiced as well. A good relationship with brokers and other scrap suppliers will often prevent problems from reaching the recycling facility in the first place.

Chapter 5 introduced the types of decoater used for decomposing the organic coatings on aluminum scrap, in particular UBCs. It was pointed out that decoating reduces melt loss of treated scrap and also reduces fluxing requirements. It also reduces soot and organic vapor generation in the melting furnace, reducing the amount of environmental equipment required for the furnace. Reducing the amount of carbon in the melting furnace off-gas may also play a key role in minimizing PCDD/F formation (Lenoir et al., 2012). By collecting the organic vapor in one process stream, postcombustion of the vapor to generate CO_2 and H_2O is easier, and the lack of flux in a decoater reduces the possibility of dioxin generation as well. Because of this, thermal pretreatment of scrap to decompose organic material is increasingly recommended for all old scrap, not just UBCs.

Decoating furnaces obviously remove water as well as organic coatings; even if decoating is not performed, a drying furnace such as that shown in Figure 14.6 is still recommended for water removal (Emes, 2004; Pluchon et al., 2011). (The heat input to the scrap in a drying furnace shortens melting times and increases productivity as well.) In addition to scrap, other inputs to a melting furnace should be kept dry— alloying elements, skimming tools, and primary or secondary ingot charged with the scrap. Water trapped in the shrinkage cavity of aluminum sows is recognized as

FIGURE 14.6 A drying furnace for aluminum scrap. (From Inductotherm, Inc., *Induction Foundry Safety Fundamentals Guide*, http://www.inductotherm.com/M2567-1108.pdf, 2008. With permission.)

one of the major causes of steam explosions (Niedling and Scherbak, 2003), and as a result preheating of any input to a melting furnace is now recommended (Jacoby, 2000). Any melting furnace with a built-in preheating system, such as a shaft or tower furnace, is useful for water removal as well.

Charging and melting: Improving safety in the charging of aluminum scrap to a melting furnace has three elements: separation of personnel from the process, placing barriers between workers and the furnace, and personal protection of workers. The separation of personnel is the most effective of the three, and the development of remotely operated charging devices is of interest in all scrap metal remelting operations (Anon, 2004; Emes, 2004; Inductotherm, Inc., 2008; Tabatabei and Turner, 2009). Continuous feeding of scrap to steel remelting furnaces has been available for some time, and equipment of this type is now being installed by aluminum remelters as well. In addition to reducing the risk of explosions and maintaining steadier furnace conditions, continuous charging seems to reduce the amount of dioxin and other organic emissions from the furnace, most likely by keeping furnace temperatures above the range in which dioxins are most likely to form (Lenoir et al., 2012; Ollenschläger and Rossel, 1999). Using conveyors or vibrating feeders rather than charge buckets also helps separate personnel from the furnace during charging (Gillespie + Powers, Inc., 2005).

Figure 14.7 illustrates a melt control station with a barrier to separate the operator from the furnace. This reduces the chance of injury from splashing of molten metal or small steam eruptions (Emes, 2004; Inductotherm, Inc., 2008). The introduction of video monitoring of charging operations now makes it possible to move the operator even farther away, and this will be more popular in the future. The aluminum industry has also made significant efforts in recent years to improve personal protection for plant personnel. These efforts include research on fire- and splash-retardant fabrics, which are more difficult to produce for molten aluminum (Johnson, 2003).

FIGURE 14.7 An operator-shielded melt control station for furnace charging. (From Inductotherm, Inc., *Induction Foundry Safety Fundamentals Guide*, http://www.inductotherm.com/M2567-1108.pdf, 2008. With permission.)

TABLE 14.3

Dioxin Concentration in Untreated and Purified Exhaust Gases from Aluminum Melting Facilities

Aluminum Melting Facility	Untreated Exhaust Gas (ng-TEQ/m³)	Treated Exhaust Gas (ng-TEQ/m³)	Treatment Method
Reverberatory furnace			Slaked lime; fabric filter
Before optimization	0.18; 1.05	0.13; 0.68	
After optimization	0.015–0.066	0.007–0.016	
Two rotary furnaces and four holding furnaces			Two-stage absorption process, followed by fabric filter
Combustion with air	42.2	9.1	
Oxy-fuel combustion	82.1	11.6	
Two rotary drying furnaces and one holding furnace (oxy-fuel combustion)	18.47	0.014	Two-stage absorption process, followed by fabric filter; slaked lime added to process

Source: Nakamura, T. et al., *Metall. Rev. MMIJ*, 17, 93, 2001. With permission.

As previously discussed, significant effort has been made in recent years to develop salt-free melting practices. This includes the introduction of vortexing and twin-chamber reverberatory furnaces, decoating equipment, and efforts to develop a salt-free dross processing technique (see Chapter 13). Eliminating salt eliminates the production of salt dross. It also minimizes chloride fumes in the plant environment and the generation of chlorinated hydrocarbons in furnace exhaust gas. For the same reasons, eliminating chlorine gas from refining operations is also a goal of many casting facilities.

Table 14.3 lists dioxin concentrations in the exhaust gas from secondary aluminum processing, again from the review of Nakamura et al. (2001). The treatment of most exhaust gas (Figure 14.8) includes (a) contacting it with slaked lime or sodium bicarbonate, which reacts with chlorinated hydrocarbons (like dioxin) to generate calcium or sodium chloride and nonhazardous hydrocarbons (Ollenschläger and Rossel, 1999), and (b) the use of a fabric-filter baghouse to recover small particles and mists from the exhaust gas (Ba et al., 2009; Lee et al., 2005). A newer approach is to quickly quench the off-gas, minimizing the time during which it is in the optimum temperature range (300°C–400°C) for dioxin formation. Activated carbon injection may also be effective at recovering PCDD/F from the off-gas (Pitea et al., 2008).

Furnace operating characteristics also play a role in dioxin formation. Firing with oxygen-enriched air tends to increase dioxin concentrations, in part by reducing the volume of exhaust gas generated (Nakamura et al., 2001). However, using dilution of off-gas with air to decrease dioxin concentration is not acceptable in many locations. The optimization of reverberatory furnace practice to preheat scrap and minimize flux usage is more helpful.

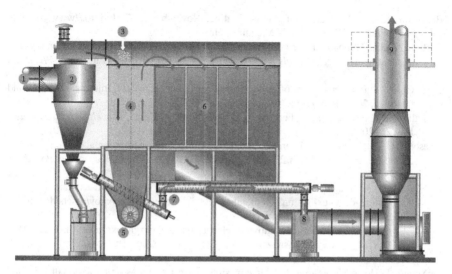

FIGURE 14.8 Off-gas handling flowsheet for secondary aluminum smelting. 1, dirty gas; 2, preseparator; 3, fresh hydrated lime; 4, reactor; 5, rotor; 6, filter; 7, re-circulation; 8, to residues silo; 9, cleaned gas. (Courtesy of Nederman Holding AB, www.nederman.se/products/ bag-filters/reverse-air-bag-dust-collectors/~/media/ExtranetDocuments/PublishedBrochure/ Primär-_und_Sekundär_Aluminiumindustrie.ashx-1.pdf. With permission.)

RECOMMENDED READING

Bertram, M.D., Hubbard, F.R., and Pierce, D.C., Scrap inspection requires ingenuity and management, in *Light Metals 2007*, Sørlie, M., Ed., TMS–AIME, Warrendale, PA, 2007, p. 797.

Jacoby, J.E., Explosions during aluminum melting in the recycling industry—Causes and prevention, in *Fourth International Symposium on the Recycling of Metals and Engineered Materials*, Stewart, D.L., Stephens, R., and Daley, J.C., Eds., TMS-AIME, Warrendale, PA, 2000, p. 869.

Pitea, D., Bortolami, M., Collna, E., Cortili, G., Franzoni, F., Lasagni, M., and Piccinelli, E., Prevention of PCDD/F formation and minimization of their emission at the stack of a secondary aluminum casting plant, *Environ. Sci. Technol.*, 42, 7476, 2008.

Pluchon, C., Riquet, J.P., Fehrenbach, F., Raynaud, G.M., Hannart, B., Jouet-Pastré, L., Bertherat, M. et al., Molten metal safety approach through a network, in *Light Metals 2011*, Lindsay, S.J., Ed., TMS-AIME, Warrendale, PA, 2011, p. 657.

REFERENCES

Anon., Automatic charging increases melting performance, *Aluminium*, 80, 390, 2004.

Ba, T., Zheng, M., Zhang, B., Liu, W., Su, G., Liu, G., and Xiao, K., Estimation and congener-specific characterization of polychlorinated naphthalene emissions from secondary nonferrous metallurgical facilities in China, *Environ. Sci. Technol.*, 44, 2441, 2010.

Ba, T., Zheng, M., Zhang, B., Liu, W., Xiao, K., and Zhang, L., Estimation and characterization of PCDD/Fs and dioxin-like PCBs from secondary copper and aluminum metallurgies in China, *Chemosphere*, 75, 1173, 2009.

BAMA, Recycling post-consumer aerosols, November 2002. http://www.bama.co.uk/pdf/ recycling_post_consumer.pdf.

Bano Shredders, UBC bales Shredding—Lattine, November 2, 2011. http://www.youtube. com/watch?v=DVfl0Vk1xUA&feature=relmfu.

Bertram, M.D., Hubbard, F.R., and Pierce, D.C., Scrap inspection requires ingenuity and management, in *Light Metals 2007*, Sørlie, M., Ed., TMS-AIME, Warrendale, PA, 2007, p. 797.

BLS (United States Department of Labor, Bureau of Labor Statistics), Injuries, illnesses, and fatalities, 2012. http://www.bls.gov/iif/oshsum.htm.

Caniglia, J., Heights engineer died trying to save others in China, *Cleveland Plain Dealer*, July 26, 2004.

Chesonis, D.C., Williams, E.M., and DeYoung, D.H., Meeting environmental challenges in the casthouse, in *Light Metals 2009*, Bearne, G., Ed., TMS-AIME, Warrendale, PA, 2009, p. 653.

CSHIB (U.S. Chemical Safety and Hazard Investigation Board), Investigation report: Aluminum dust explosion, September 2005. http://www.csb.gov/assets/document/ Hayes_Report.pdf.

Dallas Morning News, Editorial: Scrapyard relocation sets the right example for others, May 31, 2011. http://www.dallasnews.com/opinion/editorials/20110531-editorial-scrapyard-relocation-sets-example-for-others.ece.

DeYoung, D.H. and Levesque, R., Air emissions form rotary gas and rotary salt injection furnace fluxing processes, in *Aluminium Cast House Technology 2005*, Taylor, J.A., Bainbridge, I.F., and Grandfield, J.F., Eds., TMS-AIME, Warrendale, PA, 2005, p. 7.

Dougherty, P. Management of materials and wastes from salvage yard operations, February 14, 2000. http://infohouse.p2ric.org/ref/22/21186.pdf.

Ekenes, J.M., Training for preventing molten metal explosions in aluminum cast houses, in *Light Metals 2007*, Sørlie, M., Ed., TMS-AIME, Warrendale, PA, 2007, p. 807.

Emes, C.B., Designing casthouse equipment to ensure safe operation, *Alum. Int. Today,* 16(1), 35, January 2004.

Epstein, S.G., An update on the reported causes of molten metal explosions, in *Light Metals 2007*, Sørlie, M., Ed., TMS-AIME, Warrendale, PA, 2007, p. 795.

Epstein, S.G., Annual summary report on molten metal incidents, December 15, 2011. http:// www.aluminum.org/AM/CM/ContentDisplay.cfm?ContentFileID=65185&FusePreview =Yes.

Gillespie + Powers, Inc., Vibratory scrap feeder MHS, June 16, 2005. http://www. gillespiepowers.com/brochures/VBIN%20tech%20writeup.pdf.

GIZ (Deutsche Gesellschaft für International Zusammenarbeit), Introduction of a comprehensive refrigerator recycling programme in Brazil, May 2011. http://www.giz.de/Themen/ en/SID-9CCCEA81-711FBE06/dokumente/giz2011-en-proklima-projectsheet-brazil.pdf.

Glenn, K., Everything you wanted to know about MRF safety (but didn't know what to ask), *World Wastes,* 41(2), 28, February 1998.

Headland Multimedia, Industry sector: Scrap metal dealer, 2012. https://www.alliance-leicestercommercialbank.co.uk/bizguides/full/scrap/parkes-legal_matters.asp.

HSE (Health and Safety Executive, U.K.), Health and safety statistics in waste management and recycling, 2012. http://www.hse.gov.uk/waste/statistics.htm.

Inductotherm, Inc., *Induction Foundry Safety Fundamentals Guide*, August 2008. http://www. inductotherm.com/M2567-1108.pdf.

Jacoby, J.E., Explosions during aluminum melting in the recycling industry—Causes and prevention, in *Fourth International Symposium on the Recycling of Metals and Engineered Materials*, Stewart, D.L., Stephens, R., and Daley, J.C., Eds., TMS-AIME, Warrendale, PA, 2000, p. 869.

Johnson, C.D. Jr., Industry research efforts to identify FR fabrics for molten aluminum environments, in *Light Metals 2003,* Crepeau, P.N., Ed., TMS-AIME, Warrendale, PA, 2003, p. 705.

Lee, C.-C., Shih, T.-S., and Chen, H.-L., Distribution of air and serum PCDD/F levels of electric arc furnaces and secondary aluminum and copper smelters, *J. Hazard. Mater.*, 172, 1351, 2009.

Lee, W.-S., Chang-Chien, G.-P., Wang, L.-C., Lee, W.-J., Wu, K.-Y., and Tsai, P.-Y., Emissions of polychlorinated dibenzo-*p*-dioxins and dibenzofurans from stack gases of electric arc furnaces and secondary aluminum furnaces, *J. Air Waste Manage. Assoc.*, 55, 219, 2005.

Lenoir, D., Klobasa, O., Pandelova, M., Henkelmann, B., and Schramm, K.-W., Laboratory studies on formation and minimization of polychlorinated dibenzodioxins and -furans (PCDD/F) in secondary aluminium process, *Chemosphere*, 87, 998, 2012.

Léon, D.D., Casthouse safety—A focus on dust, in *Light Metals 2005*, Kvande, H., Ed., TMS-AIME, Warremdale, PA, 2005, p. 833.

Li, H.-W., Lee, W.-J., Huang, K.-L., and Chang-Chien, G.-P., Effect of raw materials on emissions of polychlorinated dibenzo-*p*-dioxins and dibenzofurans from the stack flue gases of secondary aluminum smelters, *J. Hazard Mater.*, 147, 776, 2007.

Los Angeles County, Scrap yards regulations comparison chart, March 3, 2011. http://planning. lacounty.gov/assets/upl/data/scrap_yard_comparison-chart.pdf.

Lowery, A.W. and Roberts, J., Organic coatings to prevent molten metal explosions, *Mater. Sci. Forum*, 630, 201, 2010.

Ludwig, A., Appliance disposal requires high-tech recycling, *World Wastes*, 39(6), 10, June 1996.

Malloy, J., Some opposition to proposed scrap metal yard move, July 18, 2011. http://www. trurodaily.com/News/Local/2011-07-18/article-2661762/Some-opposition-to-proposed-scrap-metal-yard-move/1.

Merz, S.K., Dioxin and furan emissions to air from secondary metallurgical processes in New Zealand, April 2004. http://www.mfe.govt.nz/publications/hazardous/dioxin-furan-emissions-vol-1/dioxin-furan-emissions-vol1-apr04.pdf.

Minter, A., India's scrap struggle, *Scrap*, 63(5), 47, November/December 2006.

Missouri Department of Natural Resources, Preventing pollution during vehicle salvage, September 2004. http://www.dnr.mo.gov/pubs/pub394.pdf.

Nakamura, T., Shibata, E., and Maeda, M., Dioxin emissions in recycling processes of metals, *Metall. Rev. MMIJ*, 17, 93, 2001.

Neitzel, R.L., Crollard, A., Dominguez, C., Stover, B., and Seixas, N.S., A mixed-methods evaluation of health and safety hazards at a scrap metal recycling facility, *Safety Sci.*, 51, 432, 2013.

Niedling, J.J. and Scherbak, M., Evaluating RSI sows for safe charging into molten metal, in *Light Metals 2003*, Crepeau, P.N., Ed., TMS-AIME, Warrendale, PA, 2003, p. 695.

Ollenschläger, I. and Rossel, H., Emission reduction in secondary aluminium industry—Investigation on charging technology, in *REWAS '99: Global Symposium on Recycling, Waste Treatment and Clean Technology*, Vol. II, Gaballah, I., Hager, J., and Solozabal, R., Eds., TMS-AIME, Warrendale, PA, 1999, p. 1015.

OSHA (Occupational Safety and Health Administration), Guidance for the identification and control of safety and health hazards in metal scrap recycling, June 5, 2008. http://www. osha.gov/Publications/OSHA3348-metal-scrap-recycling.pdf.

Pierce, D.C., Hubbard, F.R., and Bertram, M.D., Scrap melting safety—Improving, but not enough, in *Light Metals 2005*, Kvande, H., Eds., TMS-AIME, Warrendale, PA, 2005, p. 811.

Pitea, D., Bortolami, M., Collna, E., Cortili, G., Franzoni, F., Lasagni, M., and Piccinelli, E., Prevention of PCDD/F formation and minimization of their emission at the stack of a secondary aluminum casting plant, *Environ. Sci. Technol.*, 42, 7476, 2008.

Pluchon, C., Riquet, J.P., Fehrenbach, F., Raynaud, G.M., Hannart, B., Jouet-Pastré, L., Bertherat, M. et al., Molten metal safety approach through a network, in *Light Metals 2011*, Lindsay, S.J., Ed., TMS-AIME, Warrendale, PA, 2011, p. 657.

Sinkkonen, S., Paasivirta, J., Lahtiperä, M., and Vattulainen, A., Screening of halogenated aromatic compounds in some raw material lots for an aluminium recycling plant, *Environ. Int.*, 30, 363, 2004.

Sweetman, A., Keen, C., Healy, J., Ball E., and Davy, C., Occupational exposure to dioxins at UK worksites, *Ann. Occup. Hyg.*, 48, 425, 2004.

Tabatabei, E. and Turner, R.C., Molten metal splash and furnace refractory safety, *Foundry Manage. Technol.*, 137 (8), 11, August 2009.

United States Census Bureau, Industry statistics sampler: NAICS 423930, recyclable material merchant wholesalers, 2012. http://www.census.gov/econ/industry/def/d423930.htm.

Wagger, D.L., RIOS roadmap: Your guide to continual improvement in QEH&S, May 26, 2010. http://www.isrisafety.org/assets/files/presentations/RIOS.pdf.

Williams, E.M., Richter, R.T., Niedling, J.J., and Stewart, D., Preventing molten metal explosions related to skim tools and salt, in *Light Metals 2005*, Kvande, H., Ed., TMS-AIME, Warrendale, PA, 2005, p. 829.

Woodin, D., Residents oppose scrap yard permit at Joplin City Council, June 4, 2012. http://www.joplinglobe.com/local/x1647298222/Residents-oppose-scrap-yard-permit-at-Joplin-City-Council-meeting.

Index